Riffing on Strings

Riffing on Strings

Creative Writing Inspired by String Theory

Edited by Sean Miller and Shveta Verma

Scriblerus Press
New York

ISBN 978-0-9802114-0-5
ISBN 0-9802114-0-9

Scriblerus Press <http://scriblerus.net>
An imprint of Banyan Institute
548 E 82nd St., Ste. 1B
New York, NY 10028 USA

First appearances: "Desperately Seeking Superstrings?" by Paul Ginsparg and Sheldon Glashow in *Physics Today,* May 1986; "Eula Makes Up Her Mind" by Daniel Conover in *Empire of Dreams and Miracles: The Phobos Science Fiction Anthology,* eds. Orson Scott Card and Keith Olexa (New York: Phobos, 2002); "Arachne" by Elissa Malcohn in *Aboriginal Science Fiction,* Nov./Dec. 1988; "Too Many Yesterdays, Not Enough Tomorrows" by N.K. Jemisin in *Ideomancer,* Vol. 3, Issue 9, Dec. 2004; "String Theory Sutra" by Brenda Hillman in *Pieces of Air in the Epic* (Middletown, CT: Wesleyan UP, 2005); "String Theory" by Christine Klocek-Lim in *Nimrod,* Vol. 50, No. 1, Awards Issue 28, "Doing the Hundreds at 50," Fall/Winter 2006.

Cover design by Billy Bussey
Cover art "Superstring" by Félix Sorondo

Library of Congress Cataloging-in-Publication Data

Riffing on strings : creative writing inspired by string theory / edited by Sean Miller and Shveta Verma.
p. cm.
Includes bibliographical references.
ISBN 978-0-9802114-0-5 (pbk.) -- ISBN 978-0-9802114-1-2 (ebook) 1. Science--Literary collections. 2. American literature--21st century. I. Miller, Sean, 1969- II. Verma, Shveta, 1981-
PS509.S3R54 2008
810.8'0356--dc22

2008004016

To our parents, whose support and
encouragement we very much appreciate

Contents

Illustrations

Acknowledgments

We would like to thank the following people for their contributions to the development of this book.

First and foremost, we declare our appreciation for all those who contributed pieces to *Riffing on Strings*—you are a remarkable group of talented writers who represent a richly diverse range of styles and ideas.

Our parents Kiran and Dalbir Verma, Jeff Miller and Susan Sealy have been unwavering in their support. We whole-heartedly appreciate everything they've done for us.

The idea for this book came about on a vacation we took in August of 2006 at the wonderful home of Shveta's cousin, Vikas Bhushan. The beautiful beachside environs of Paia, Maui were instrumental in unclogging creative pathways. We want to thank Vikas for being such a gracious host on that holiday—and the year after, when he and the house played host to us and kin once again, this time for our wedding.

Sean would like to thank Andrew Gibson, who for the past two and a half years has served as his PhD thesis supervisor at Royal Holloway College. Andrew's expert oversight of Sean's thesis has been highly motivating. While the looming gauntlet of thesis submission and viva promises to be nerve-wracking, Sean has every confidence it will be rendered manageable by his supervisor's guidance.

Sean would like to thank Laura Salisbury for serving so well as his masters dissertation supervisor at Birkbeck College, University of London—and for her ongoing friendship. He would also like to thank Steven Connor for giving a lecture at the college that provided the initial spark for this book.

We also want to thank Michael-Cade Stewart and Matthew Wraith, who were both classmates of ours on the Birkbeck course, for their company when Sean is in London and for their highly engaging conversation. Sean particularly appreciates Matt's having hosted him many times in his flat—and Michael's volunteering to comment on the introduction and, generally, to help out in other ways.

We also want to thank Hayley Holness and James Harrigan for their easy-going hospitality in London, as well as for numerous and always stimulating discussions on topics ranging over everything under the sun. Our thanks also go out to Alf Metelius and Catherine Harrington for their thoughtful reading of an early draft of the introduction.

We also owe a debt of gratitude to all those who submitted work to the collection. Due to considerations of space, we had to refuse many truly interesting pieces. We appreciate your efforts.

Introduction

Sean Miller

Like many people, my first encounter with string theory was through Brian Greene's book *The Elegant Universe*, which came out in the U.S. in 1999. I had always been interested in physics. In particular, I was intrigued by the physics that treated the peripheries of our capacity to experience the world—the realm of the incredibly small, the microcosmos, and the realm of the mind-bogglingly large, the macrocosmos in all its awesome vastness. And here was string theory, a scientific theory that promised to knit these two seemingly disparate worlds into one integral whole. A theory of the tiny and the humongous rolled up into one neat, elegant formalism. It is for this reason that string theory has been called a theory of everything. Little did I know then, as I tussled enthusiastically with the ideas in *The Elegant Universe*, that it would set me on a trajectory that I still find myself on to this day.

Physicists mean something specific when they speak of a theory of everything. They certainly do not mean to suggest that string theory can explain why sunflowers lean to the light or why people fall in love. What they mean by a theory of everything is this: string theory holds the potential to reconcile the two great high energy physics theories of the past century into one consistent mathematical formalism. Those two theories are quantum theory, the reigning theory for explaining the workings of the realm of the very small, and Albert Einstein's theory of general relativity, which explains with great precision the realm of the vast. General relativity is a theory of gravity, of how massive objects attract each other—it pertains to things like cannon balls, rockets, planets, stars, galaxies, galaxy clusters, black holes, and the entire universe itself.

In the early part of the twentieth century, quantum theorists found that the atom is, in fact, divisible. It is made up of a nucleus and an electron "shell" or "cloud" that orbits the nucleus. As you may recall from high school physics, the atomic nucleus is made up of subatomic particles called protons and neutrons. In the seventies, quantum theorists discovered that these protons and neutrons aren't fundamental either—they too are made up of smaller parts, what Nobel Laureate Murray Gell-Mann whimsically called "quarks," from a line in James Joyce's *Finnegans Wake*. Quarks and electrons—the fundamental ingredients of all matter. Quantum theory has enjoyed unparalleled experimental success. And yet physicists find it unsatisfactorily incomplete. Most glaringly, it does not incorporate gravity.

Enter the string. String theory claims that all the subatomic particles described by quantum theory—the quarks and electrons that make up matter, as well as all the subatomic particles that express force—are actually different vibrations of a more fundamental object, the string. Strings can be either closed like a loop or open like a wiggly filament. A string only looks like a particle when you view it from a distance, much in the same way that a house, an extended object, looks like a dot when gazing down from the window of a jet at 30,000 feet. The most powerful instruments today—that are able to peer into the microcosmos—can only probe distances of a few orders of magnitude below the scale of the atomic nucleus. That is why, according to string theorists, quarks and electrons appear to us as points. Strings, on the other hand, are on the order of 10^{-34} meters, what string theorists call the Planck length, named after the famous German Nobel Laureate Max Planck. That's a hundred billion billion billion times smaller than a speck of dust! Brian Greene writes: "To get a sense of scale, if we were to magnify an atom to the size of the known universe, the Planck length would barely expand to the height of an average tree."

Physical theories such as string theory deal with the fundamental objects that constitute our world and how they interact with each other. Matter either attracts or repels other matter. This attraction or repulsion at different scales and under different conditions is what physicists call force. As you know, gravity is a force of attraction. But gravity is very weak at atomic scales. It is only on scales of cannon balls, jets, and stars—where much matter has accumulated—that the force of gravity plays a significant role. Physicists have also nailed down three other forces that come into play on atomic scales. They are electromagnetism, the strong nuclear force, and the weak nuclear force.

Quantum theory describes accurately these three other forces. Electromagnetism, of course, can either attract or repel matter, depending on the charge, negative or positive, of the matter involved. The strong nuclear force binds quarks together in the nuclei of atoms. It is an attractive force. The weak nuclear force repels subatomic particles under certain conditions—and is responsible for certain kinds of radioactive decay. Quantum theory is incredibly accurate at describing the interaction of matter on atomic scales in terms of these three fundamental forces.

General relativity is highly effective at describing gravity on large scales. It essentially says that matter is a form of energy and that that energy constitutes space and time itself. Furthermore, space and time warp around massive objects, such as stars, because of the energy manifest in those objects. In extreme cases, such as within a black hole, the force of gravity is so powerful—that is, the amount of matter present is so vast—that even light cannot escape its gravitational pull. Yet general relativity is inadequate when it comes to describing black holes because a black hole is not only very heavy, it is also very small. All that matter has been compacted down to atomic scales. This represents a situation where quantum theory should be applied.

But when theorists try to apply both quantum theory and general relativity to natural phenomena that are both incredibly dense and incredibly small—such as black holes and the very early universe before its explosive expansion—both theories give nonsensical answers. String theory, because it incorporates consistently the force of gravity, as well as the three fundamental forces of quantum theory, promises to overcome this conflict. It is a quantum theory of gravity. This is what gets string theorists so excited about its prospects.

A few years back, I did a Masters course in literature at Birkbeck College, University of London. In the spring, Steven Connor gave a lecture that served as the spark that ignited my growing interest in string theory as a cultural phenomenon. At the risk of oversimplifying, I want to share with you the observation that Connor made that, in many respects, set me on my current path of research on the "cultural currency of string theory." In the talk, Connor spoke of the so-called Two Cultures, an expression made famous by the physicist and novelist C.P. Snow in 1959 to describe the breakdown in communication between those who worked in the sciences and those in the humanities. One consequence of this chasm was a tendency for humanities types—artists, writers, scholars—to make use of analogies and metaphors that were based on anachronistic scientific notions. Our imaginations—the underlying images and ideas that inform the structure of our thinking—are, in certain respects, still stuck on nineteenth-century scientific conceptions of the world, a world largely understood from a Newtonian perspective as mechanistic and deterministic, and where time advances linearly from past to present to future.

With Greene's *The Elegant Universe* buzzing fuzzily in my memory, I asked Professor Connor whether a contemporary physical theory such as string theory might offer us humanities types a perspective from which to re-imagine the social world that concerned us. He replied, yes, absolutely. Not one needing an excess of encouragement to fly off on a speculative tangent, my imagination grew increasingly fixated in the months that followed on this idea of how string theory as a worldview might influence contemporary culture in novel and interesting ways.

Of course, such flights of fancy come with a risk. One of the consequences of the rift between the Two Cultures is that when we in the humanities do evoke ideas from the so-called hard sciences, like string theory, those in the hard sciences tend to look upon our efforts with a certain derision. Humanities types, in turn, tend to view scientists with suspicion as well—as reductionists oblivious to the subtleties of language and culture. The chasm between the Two Cultures, this lack of understanding, has led to a general mood of mutual animosity and distrust.

One poignant example of this animosity is the notorious "Sokal Affair" of the mid-nineties. In 1996, Alan Sokal, a physicist at New York University, submitted a paper entitled, "Transgressing the Boundaries: Towards a Transformative Hermeneutics of Quantum Gravity," to what was then a prestigious sociology journal called *Social Text*. At the time, *Social Text* was at the forefront of what is called in The Strong Program, an effort by postmodern sociologists to discredit the "objectivity" of scientific discourse.

Sokal intentionally riddled the paper with pseudo-scientific gibberish in order to back the claim that, as he puts it, "In quantum gravity [...] the space-time manifold ceases to exist as an objective physical reality; geometry becomes relational and contextual; and the foundational conceptual categories of prior science—among them, existence itself—become problematized and relativized. This conceptual revolution, I will argue, has profound implications for the content of a future postmodern and liberatory science." Much to Sokal's surprise and delight, the editors at *Social Text* failed to vet the paper for its scientific inaccuracy and went ahead and published it. When Sokal revealed the paper to be a hoax, chock as it was with outrageous abuses of theoretical physics jargon, a controversy was born.

In a follow up book co-authored with Jean Bricmont called *Fashionable Nonsense: Postmodern Intellectuals' Abuse of Science*, Sokal writes: "When concepts from mathematics or physics are invoked in another domain of study, some argument ought to be given to justify their relevance." Of course, physicists like Sokal and Greene wouldn't place the same restrictions on other forms of artistic expression, such as science fiction. Safely in the realm of play, poetry and fiction are free to make use of ideas from hard science in whatever whimsical way they see fit. This is because they make no claims on the objectivity of the "real world." They deal with feelings, the imagination, and other "subjective" human phenomena.

This tidy sorting of subjectivity from objectivity is typical of those in the hard sciences, schooled as they have been in the dual philosophies of positivism and reductionism. Positivism and reductionism have had a profound influence in shaping the perspectives of the hard sciences such as theoretical physics. Simply put, positivism argues that what you cannot directly confirm or deny through experiment is not worth bothering with. From this perspective, any investigation of such things can only be idle speculation concerned with abstractions largely irrelevant to the real world. It is not physics, but metaphysics. As a complement to positivism, reductionism argues that complex worldly phenomena can best be understood by reducing them to their smallest and simplest constituent parts and understanding the rules of their interaction.

My own research tries to show that, with respect to string theory, this supposedly neat distinction between the imagination and the objective world is not so neat after all. And it is something that I believe the pieces in this collection demonstrate as well. In that spirit, the question central to *Riffing on Strings* is: what can we say and feel about the cosmos we live in through the prism of string theory imaginaries?

A Brief Tour of String Theory

For those of you who are unfamiliar with string theory, I offer you a brief tour of its most salient features—within the context of its history. If you're up-to-date on the topic, you may want to skip ahead to the next section of this introduction. Keep in mind, though, that the account that follows may add a few new twists to the string theory story that are well worth considering.

In its essence, the history of string theory is the story of people with highly specialized training working together to solve difficult problems within a well-defined field of activity. String theory began in the late nineteen-sixties, when theoretical physicists were trying to come to terms with the mysteries of the atomic nucleus. At the time, results from experiments conducted at the particle colliders physicists use to probe the structure of the subatomic world suggested that protons and neutrons were not, in fact, fundamental. In quantum theory, protons and neutrons belong to a class of subatomic particles called *hadrons*, a word which comes from the Greek *hadr-*, meaning heavy. They are the massive particles that, among other things, make up the nuclei of atoms. When experimentalists in the sixties smashed atoms together, they found that hadrons behaved in perplexing ways—as they flew further apart, the force between them seemed to get stronger.

At the time, there were several hypotheses proposed to explain this odd behavior. One was Richard Feynman's theory of *partons*. Another was Murray Gell-Mann's theory of *quarks*. Both of these theories, while significantly different in detail, took a reductionist approach. They argued that protons and neutrons could be explained in terms of smaller, constituent parts. Yet another was Geoffrey Chew's theory of the S-matrix. Unlike the parton and quark theories, Chew claimed that his approach was holistic. Simply put, it was the whole system, the S-matrix, out of which the parts, hadrons, emerged. Chew's theory was very fashionable among physicists at the time.

Then, in string theory lore, in 1968 along came an Italian postdoctoral fellow at CERN in Geneva, Switzerland named Gabrielle Veneziano. Veneziano was a disciple of Chew's S-matrix theory and was eager to make his reputation in the field. While scouring through an antique mathematical text authored by the eighteenth-century mathematician Leonhard Euler, he came across a formula, called a Beta function, which seemed like it could be used to model the strange behavior of hadrons. No one had a convincing explanation as to why the force that held hadrons together grew stronger the further the hadrons moved apart, much like a rubber band. The Beta function mathematically described elastic behavior like this. Veneziano published his idea in a professional journal and it caused a stir. Other theorists began to work on the problem using Veneziano's approach.

A couple of years later, Yoichiro Nambu, Leonard Susskind, and Holger Nielsen, all working independently, came out with papers that argued that the force that held hadrons together in the nuclei of atoms could best be described not as point-particles, as quantum theory stated, but as objects extending along an extra dimension. While all three used the expression "harmonic oscillator" to describe this strange new theoretical object, each one imagined it a bit differently. Nambu called it a "cavity resonator." A cavity resonator is an enclosed space where energetic excitations produce harmonic oscillations. Flutes and organ pipes are examples of cavity resonators—as is the body of a violin. Nielsen called his model a "fishnet diagram," where each strand of the fishnet represents one particular pattern of oscillation. Susskind likened his model to a "chain or springs" or "an ideal rubber band." But ultimately, it was Susskind, in another paper a year later, who named this new one-dimensional extended object a string. Like a rubber band, a string has tension. When plucked, that is, when it interacts energetically with other strings, it vibrates. The various modes of vibration—or to use a musical analogy, the notes—of a string precisely describe the scattering patterns of hadrons as they ricochet off each other.

Susskind tells the story of how the name stuck. Shortly after the publication of this second paper, he went to a big conference in Coral

Gables, Florida. Susskind suddenly found himself in an elevator with the great Murray Gell-Mann, who asked him what he did. Susskind replied, "I'm working on this theory that hadrons are like rubber bands, these one-dimensional stringy things." Gell-Mann began to laugh and laugh. Susskind writes:

> I didn't see Murray again for two years. Then, there was a very big conference at FermiLab, and a thousand people were there. And me, I'm still a relative nobody. And Murray is in constant competition with his colleague Richard Feynman over who is the world's greatest physicist.
>
> As I'm standing there talking to a group of friends, Murray walks by and in an instant turns my career and my life around. He interrupts the conversation, and in front of all my friends and closest colleagues, says [...] "The stuff you're doing is the greatest stuff in the world. It's absolutely fantastic, and in my concluding talk at the conference I'm going to talk about nothing but your stuff."
>
> On the last day of the conference, off in the corner somewhere [...] the next thing I heard was Murray holding forth. He was telling a group of his cronies everything I had told him. "Susskind says this, and Susskind says that. We have to learn Susskind's String Theory."

Still more theoretical physicists became interested in the topic, which led to a flurry of advances in this new "string theory" throughout the early seventies. However, in the meantime, Gell-Mann's theory of quarks was also making great strides. Improvements in collider technologies allowed experimentalists to probe the nucleus at smaller and smaller scales. Results from these experiments strongly corroborated Gell-Mann's quark theory over Susskind's string theory. Gell-Mann's theory of quarks has since become known as *quantum chromodynamics*, the theory of the strong nuclear force, which binds together atomic nuclei (it is a "chromodynamics" because different kinds of quarks have different "colors"—not actual colors but whimsically so). Quantum chromodynamics is a point-particle theory that describes the binding of the strong nuclear force in terms of the exchange of special particles called gluons (as in "glue") between sets of quarks. Joined with the theories of electromagnetism (*quantum electrodynamics* or *QED*) and the weak nuclear force (*electroweak theory*), the three fundamental forces have since come to be known as the Standard Model.

With the success of quantum chromodynamics in the mid-seventies, most physicists abandoned string theory. A small group of mavericks continued to tinker with it, though, hoping to modify string theory in order to make it more realistic. Instead of modeling hadrons, their strategy was to use string theory to describe not just the strong nuclear force, but the other two subatomic forces as well. This was known as *bosonic string theory*.

In the Standard Model, there are two main classes of subatomic particles, *bosons*, named after the Indian physicist Satyendra Nath Bose, and *fermions*, named after the Italian Nobel Laureate Enrico Fermi. All particles have what is called spin—a mathematical attribute like the spinning of a top. (A point particle, since it has no extent, cannot actually spin.) Bosons, most of which are considered force particles, have a spin of one. They return to their original state after one rotation. Fermions, most of which are considered matter, have a spin of one-half. They return to their original state after two rotations. In the early seventies, theorists succeeded in finding a way to use string theory to describe all the bosons in the Standard Model. They also made the theory consistent with Einstein's theory of special relativity, an important part of general relativity which demands that no particles may travel faster than the speed of light.

There was one striking consequence in reformulating string theory to model bosons. In order to be mathematically consistent, the tiny vibrating strings of energy had to exist in 26 dimensions of spacetime—that is, in one dimension of time and twenty-five of space! Our bodies move in three dimensions of space—back and forth, up and down, and left and right. We understand the universe in its entirety as having four spacetime dimensions, three of space and one of time. Needless to say, it is difficult to comprehend the meaning of 25 dimensions of space. This need for 26 spacetime dimensions in bosonic string theory was met with a great deal of skepticism from the physics community.

In 1975, theorists Joel Scherk and John Schwarz published a paper that extended string theory to incorporate not only bosons, but fermions as well. This new version of string theory could now generate a close approximation of the entire spectrum of particles in the Standard Model. But by including fermions with bosons, Scherk and Shwarz found that, in order to be mathematically consistent, the strings had to reside not in 26, but 10 spacetime dimensions. One surprising result of this synthesis was that a set of string vibrations for gravity naturally emerged from their model. Because string theory naturally included gravity, they reasoned that it may very well be a legitimate quantum theory of gravity. This was the move that inspired some physicists to believe that perhaps string theory could indeed be a theory of everything. Another result of Scherk and Schwarz's new version of string theory was that the scale of the string shrank from the scale of hadrons, about 10^{-16} centimeters, to the Planck scale, or 10^{-33} centimeters. Yet there were still several nagging inconsistencies in Scherk and Schwarz's theory of strings for bosons, fermions, and gravity.

These inconsistencies were ultimately resolved with the incorporation of yet another theory—*supersymmetry theory*. In a nutshell, supersymmetry provides a way for bosons and fermions to transform into one another. In quantum theory, the various classes of particles are organized into three groups of what are called gauge transformations—ways that these particles may change when they interact with other particles. The expression "gauge" comes from nineteenth-century railroad terminology, where different track widths were classified as gauges. There is a gauge group in quantum theory for each of the three fundamental forces: the electromagnetic gauge, the strong gauge, and the weak gauge.

In quantum theory, there is no obvious connection between these three forces and their corresponding gauge groups. They are simply cobbled together as a kind of hodgepodge, each pertaining to its own relevant situation. Supersymmetry promises to unite these three disparate gauge groups into one supergroup. This supergroup allows for certain rigidly defined transformations amongst all its members. In a supersymmetric world, a boson may transform into a fermion and a fermion may change into a boson. What this means is that each boson in the Standard Model has a superpartner fermion. Accordingly, each fermion has a superpartner boson. No one has ever observed these superpartners in nature, but one of the goals of the new collider going online in 2008 at CERN, the Large Hadron Collider, is to search for these superpartners.

While supersymmetry theory is independent of string theory—it has also been incorporated into current versions of point-particle quantum theories—the discovery of such superpartners in nature would help string theory's case considerably. This is because supersymmetry complements string theory in elegant and precise ways. The first to successfully incorporate supersymmetry into string theory were the theorists F. Gliozzi, J. Scherk, and D. Olive in 1976. But shortly afterwards, theorists faced a number of intractable problems within this exciting new *superstring theory* that prompted other physicists to disregard it, especially in light of the continued success of the Standard Model. Two physicists who persisted, against the odds, in their work on superstring theory were John Schwarz and Michael Green. In 1981, Green and Schwarz were able to make superstring theory consistent with special relativity. In 1984, they were able to resolve a nagging issue with what are called "anomalies," mathematical inconsistencies that previously rendered their theory hopelessly unrealistic. Most string theorists consider 1984 a watershed moment in the ascendancy and subsequent institutionalization of superstring theory. Their solution to these substantial problems in superstring theory made Green and Schwarz instant celebrities within the physics community.

In the wake of their success, many more talented physicists joined the growing flotilla of those working on superstring theory. There was a widespread sense that a breakthrough of historical proportions was teasingly just over the horizon. Out of this flurry of intensive research came four alternative versions of superstring theory. In 1985, David Gross, Jeffrey Harvey, Emil Martinec, and Ryan Rohm, all working at Princeton University, proposed what has come to be recognized as the most successful of these anomaly-free superstring theories in reproducing a "semi-realistic" quantum particle spectrum: Heterotic $E_8 \times E_8$ Superstring Theory. The odd thing about heterotic superstring theory, though, is that unlike the others, which include 10 spacetime dimensions, it couples a gauge group that requires 10 spacetime dimensions to one of 26 spacetime dimensions. This is what makes it "heterotic," from the Greek meaning "different." It strikes me as ironic that the most "realistic" superstring theory, the one that supposedly most resembles our universe, is also the weirdest in terms of its mathematical structure.

The five versions of superstring theory differed enough to pose a serious problem: which one was the correct one? It was an embarrassment of riches. Most physicists feel that a theory should carry with it a sense of inevitability. Inevitability requires that only one unique theory ought to be the correct one and that its correctness can be borne out by its making specific, testable predictions that other theories do not. None of the five competing superstring theories proposed in the mid-eighties suggested such an inevitability.

The confusion that ensued over this embarrassment of versions was further exacerbated by the problem of what to do with the six extra spatial dimensions. At best, superstring theories can only produce what theorists call "semi-realistic" physics. By "semi-realistic" they mean a given theory can produce a particle spectrum that closely resembles the particle spectrum of the Standard Model but not exactly. There are loose ends, gaps, extra elements, niggling inconsistencies. One strategy for making these superstring theories realistic is to tinker with the geometry of the extra spatial dimensions. It is assumed that since experimental equipment cannot detect these extra dimensions, they must be small, on the order of the Planck scale.

As a consequence, in the late eighties, theorists concerned themselves with formulating the various superstring theories where these extra dimensions were compactified—made tiny and bundled together. At the time, the issue was how to describe these ultramicroscopic bundles of extra dimensions. String theorists found some success using a funky geometrical object that had been explored by an Italian-American mathematician, Eugenio Calabi, back in the late 1950s. Calabi's conjecture was subsequently proved in 1977 by the Chinese mathematician, Shing-Tung Yau. In honor of their discoverers, these balled-up, contorted, knot-like multidimensional objects are called *Calabi-Yau manifolds* (also sometimes referred to as Calabi-Yau "shapes" or "spaces").

While Calabi-Yau manifolds tantalized string theorists with the prospect of the perfect shape to generate a realistic particle spectrum, finding the correct one proved to be elusive. There are millions upon millions of potential candidates—many of them produce semi-realistic physics—but none has been shown to produce the exact solution. To compound the problem, it is a painstaking process to sift through all the promising candidates. Theorists have also fiddled with comparable multidimensional geometric objects called *orbifolds* and *orientifolds*, but, as of this writing, the verdict is still out.

So not only are there five competing versions of superstring theory to contend with, there are millions of potential ways to configure the extra-dimensional spaces within these theories. In effect, there are millions of superstring theories to choose from, none of which is more than, at best, semi-realistic.

Then in 1995, at the annual Strings conference, Edward Witten proposed a striking new passage through this theoretical impasse. In a lecture that shocked and enthralled his audience, Witten argued that the five superstring theories may, in fact, be different formulations of a more fundamental theory, which he later dubbed *M-theory*. Witten has said that the "M stands for magic, mystery or membrane, according to taste." Others have suggested that the M stands for "matrix," since the theory makes use of infinite-dimensional matrices. Still others suggest that the M stands for "mother" as in "the mother of all theories." The more skeptical have offered yet more decipherings of the M in M-theory. It might be an upside-down W, and thus a small gesture of self-aggrandizement on Witten's part. Or it might as well stand for "murky" or "monstrous," since M-theory promises much but actually resolves little. Theorist Michael Duff has facetiously called M-theory "the theory formerly known as strings."

One notable feature of Witten's M-theory is that by teasing out the equivalences among the various superstring theories, spacetime winds up having not 10 spacetime dimensions, but 11. Another feature of M-theory is that strings, the supposed fundamental objects of the universe, take a back seat to the brane, short for membrane. While strings are extended one-dimensional vibrating objects, branes are two or more dimensional extended objects. A useful analogy is to imagine a violin string. When a violin string is plucked, it vibrates at a certain frequency, depending on the length of the string. We hear this as a particular note. Now imagine a brane as a drumskin. When a drumskin is hit, it also vibrates at a certain frequency, a particular "note." We can now extrapolate this analogy by adding additional dimensions to the brane.

In M-theory, the universe is composed of a variety of different branes, categorized by the number of dimensions they possess. A 0-brane is a point-particle-like object. A 1-brane is a particular kind of string. A 2-brane is a two dimensional membrane, like a drumskin. A 3-brane has three spatial dimensions, and so on, up to a 10-brane. (Remember that one dimension corresponds to what we experience as time.) Michio Kaku's essay in this collection, "M-Theory: The Mother of all Superstrings," goes into more detail about the importance of branes in M-theory.

Most recently, theorists have been calling into question the assumption that these extra spatial dimensions in M-theory must be compactified. There have been some suggestions that the extra dimensions, while still beyond the reach of our instruments, are larger than the Planck scale, and therefore may be just tantalizingly out of reach. Another line of inquiry in current research explores the possibility that the extra dimensions may be extremely large, on cosmic scales. These models are often called "braneworld scenarios." In some of them, our four-dimensional universe is situated on a 3-brane that, in turn, resides within an eleven-dimensional meta-verse, often called "the bulk."

One peculiar version of the braneworld scenario argues that there is a nearly infinite number (10^{500} or thereabouts) of such braneworlds in the bulk. Each of these braneworlds has laws of physics that vary from its neighbors. That variation, however minute, alters the basic composition of matter and force within the braneworlds. To explain why we in our universe find the particular balance of matter and force that would allow for our very existence as human beings, such models evoke what is called the Anthropic Principle. Physicist P.C.W. Davies defines the Anthropic Principle thus: "we should observe a universe of minimal order consistent with the existence of observers." The Anthropic Principle suggests that the very fact that we humans are here demonstrates that the universe we live in must be finely tuned. That is, the laws of physics must be such that they generate the particular balance of matter and force that results in the stars, planets, chemistry, and ultimately the complex biology which we as human beings represent. For if our universe were not so finely tuned, we could not have come about.

The question then becomes: how is it that our universe is so finely tuned? Landscape Theory, as it is called, answers this question by suggesting that it is highly plausible that just such a universe as ours is bound to come about in a bulk that contains 10^{500} braneworld universes. With that kind of variety, our particular universe—finely-tuned as it is—is only special in that it is the one that happens to have produced us.

As a matter of convenience, I will refer to this panoply of theories under the umbrella term "string theory," even though, strictly speaking, there is no one correct and inevitable string theory *per se*. Shuffling all these string-related theories into the bucket category of string theory is simply a matter of rhetorical convenience.

To briefly review: according to string theory, the universe is composed of certain basic ingredients. They are the string, the brane, and extra dimensions. The string is a one-dimensional extended object that vibrates. It can be open, like a filament, or closed, like a loop. A string has tension. The amount of tension in the string determines at what frequency the string resonates, and as a consequence, what quantum particle the string becomes. The various resonance modes of the string generate the entire particle spectrum—all quantum particles of matter and force, including gravity.

String theory also contains another fundamental object called the brane. A brane is a two or more dimensional extended object. Like strings, a brane has tension and vibrates in a certain range of frequencies. This set of resonances determines the behavior of the brane in relation to other branes (and strings).

We traditionally understand our universe to have three dimensions of space and one of time. But in order for string theory to be mathematically consistent, the universe must have either 26, 10, or 11 spacetime dimensions, depending on which version of string theory we consider. Bosonic string theory needs 26 spacetime dimensions in order to be mathematically consistent. Superstring theory needs 10, and M-theory needs 11. Some string theory-based models compactify the extra dimensions into various knot-like geometric structures, such as Calabi-Yau manifolds. Other models extend the extra dimensions out to cosmic scales.

There is a great deal of controversy in the physics community over whether these extra dimensions are real or are actually just mathematical artifacts. And since strings, branes, and extra dimensions reside on scales impossibly remote from the reach of our scientific instruments, we may never hope to verify their existence through experiment. Peter Woit, whose piece in this collection pokes subtle fun at these and other "metaphysical" aspects of string theory, echoes the famous quantum theorist Wolfgang Pauli when he suggests that string theory is "not even wrong." The essay "Desperately Seeking Superstrings?" by Nobel Laureate Sheldon Glashow and ArXiv.org creator Paul Ginsparg also bitingly critiques string theory as physics gone astray.

String Theory as Scientific Imaginary

What I have given here is a whirlwind tour of string theory that does not do justice to its subtleties. If you are hungry for a more thorough explanation of string theory, I've provided a list of suggested readings at the back of the book. As an object of study, string theory is deep and multi-faceted. You can approach it from a host of perspectives.

A full appreciation of string theory must contend with the controversy surrounding its legitimacy as science. Does string theory simply reduce to frivolous conjecture and mathematical pyrotechnics? Is it satisfying to accept that extra dimensions are hidden because of our current technological limitations, or are they merely misleading figments of the imagination? Will string theory ultimately provide humanity with a theory of everything, one "master equation," as Brian Greene puts it, out of which all the mysterious workings of the universe are laid bare?

Adding flavor to these still as yet unresolved questions are the curious details of string theory's history. Ed Witten has said that "superstring theory is a piece of twenty-first-century physics that fell by chance in the twentieth century." In many respects, the problems of string theory are so obdurate that not only can no one solve them, but theorists simply lack the mathematical tools to do so. These tools have yet to be invented. As such, if Gabrielle Veneziano hadn't stumbled upon that antique formula of Leonhard Euler's back in 1968, would there even be a string theory as we know it today? And now that we find ourselves firmly planted in the twenty-first century, will our mathematics catch up with string theory's various mysteries?

These questions are further complicated by how theoretical physics itself gets produced. There are currently about a thousand string theorists engaged in active research in the world. It takes many years of training to gain the skills necessary to participate in the string theory technical conversation. String theory as a professional practice is an exclusive coterie. Even those who work in related disciplines, such as quantum field theory or general relativity, have a hard time understanding its argot. Lee

Smolin of the Perimeter Institute, who specializes in an alternative to string theory called "loop quantum gravity," writes: "Even now, one can go to a conference and find that string theory and loop quantum gravity are the subjects of separate parallel sessions. The fact that the same problems are being addressed in the two sessions is noticed only by the small handful of us who do our best to be in both rooms." In some respects, an inevitable consequence of the advancement of scientific knowledge is its ever-increasing fracturing through specialization.

Since the mid-1990s, string theorists have been making a concerted effort to share their work with a wider, non-specialist audience, perhaps out of concern for their relative professional (and theoretical) isolation. The first attempts to popularize string theory to a lay audience occurred in 1987 with Michio Kaku and Jennifer Trainer's *Beyond Einstein* and P.C.W. Davies and Julian R. Brown's *Superstrings: A Theory of Everything?*. Since then, popular accounts of string theory have been coming out with increasing regularity. With the success of such works as Kaku's *Hyperspace* in 1994 and Greene's *The Elegant Universe* in 1999, string theory has thoroughly infiltrated public awareness.

References to string theory pop up in the oddest of places. For instance, in the debut season (2007) of the CBS sitcom *The Big Bang*, its two main characters, both young male physicist nerds, debate, in a glib fashion typical of primetime TV, the merits of "bosonic string theory." They do this to impress a female colleague whom one of them hopes to seduce, however clumsily. There is a trilogy of *Star Trek Voyager* novels subtitled *String Theory*. In the Wikipedia entry for string theory, a section towards the bottom of the page lists pop culture references. Needless to say, these references and more extended treatments of string theory continue to multiply.

These are, of course, isolated examples, but they serve to highlight what you might call the second phase in the dissemination of scientific ideas. The first phase is didactic. This is where scientists themselves—sometimes with the help of journalists—introduce their disciplines and discoveries to a non-professional audience. Through such efforts, scientists hope to educate non-specialists for more or less three reasons: one, from a genuine desire to share their enthusiasm for their subject; two, in order to recruit young people to the profession; and three, to gain enough popular support for their research such that it will eventually contribute to continued public funding, since most scientists work for universities that depend on both government grants and alumni support.

In the second phase, the key ideas of a scientific discipline enter the wider popular dialogue as buzz words. These ideas are dropped in passing—around office coolers or at cocktail parties—as a kind of status symbol that shows those who evoke them to be savvy consumers of the cutting-edge. Lisa Randall writes:

> I realized how much attention [M-theory] was receiving when I was on a plane returning from London. A fellow passenger, who turned out to be a rock musician, saw that I was reading some physics papers. He came over and asked me whether the universe had ten or eleven dimensions. I was a little surprised. But I did answer and explained that in some sense, it is both. Since the ten- and eleven-dimensional theories are equivalent, either one can be considered correct.

But with respect to this collection, it is the third phase in the dissemination of string theory as a "scientific" idea that most interests us. This is where the images associated with a scientific theory become detached from that theory's formal expression and take on a life of their own. In the case of string theory, its formal expression is couched entirely in the language of mathematics. To understand the mathematics of string theory, you need to be a professional. There are precious few non-professionals in the world who can say with confidence that they can read and understand the mathematical arguments made in string theory technical discourse. Those arguments must both be understood in their own terms—the consistency of the mathematics itself—and in terms of the broader context of accepted physical theory and potential experimental evidence to support the claims of those arguments. A science such as string theory must, in this sense, corroborate the successes of its predecessors—quantum theory and general relativity—while also shedding new light on heretofore unresolved mathematical and evidential inconsistencies in those predecessors.

But culture at large—a lay audience that includes other scientists who do not specialize in string theory—has no access to the formal conversation of string theory. We only have access to its exposition. Playing off the etymological root of the word "exposition," as a lay audience, what we get with string theory is that which has been "put out" or exposed to us. It is a string theory mediated through expository prose and, most significantly, the imagery that constitutes that prose.

We are accustomed to thinking of concepts as something categorically distinct from imagery. Mathematical arguments are surely an extremely sophisticated form of conceptualization. But what we often don't appreciate is that the concepts we make use of in ordinary expression—in the prose of popular accounts of science, for example—come riddled with images. These images do not merely adorn concepts. They are not tossed into the mix merely to make those concepts more "clear" and accessible. Images are, in fact, fundamental to exposition—and to the conceptualization that exposition represents. I challenge you to think of a concept that, at its root, does not arrive as a shell with some image tucked within it. Reason depends integrally on the imagination. The two cannot be purified of each other.

An impassioned defense of this position, which might strike some of you as far-fetched, is beyond the scope of this introduction. As a proxy, I offer you this passage from linguists George Lakoff and Mark Turner's book, *More Than Cool Reason*. It addresses this classic debate between old school literalists and second-generation cognitive scientists (in reading this excerpt, remember that metaphors are made of images):

> The Literal Meaning Theory entails [...] the assumption that reason and imagination are mutually exclusive. Reason is taken to be the rational linking up of concepts, which are nonmetaphoric, so as to lead from true premises to true conclusions. Thus, there is nothing metaphoric about reason, neither its operations nor the concepts it operates on. Metaphoric reasoning, on this view, cannot exist. Since metaphor is excluded from the domain of reason, it is left for the domain of imagination, which is assumed to be fanciful and irrational. This view is, as we have seen repeatedly, erroneous. Many of our inferences are metaphoric: we often reason *metaphorically,* as when we conclude that if John has lost direction, then he has not yet reached his goal. Our reasoning that time changes things is metaphoric and deeply indispensable to how we think about events in the physical, social, and biological worlds. Indeed, so much of our reason is metaphoric that if we view metaphor as part of the faculty of the imagination, then reason is mostly if not entirely imaginative in character.

In keeping with this perspective, let's examine the image of the string in string theory. Here is a formula from bosonic string theory called the Nambu-Goto Action:

$$S = -\frac{T_0}{c} \int_{\tau_i}^{\tau_f} d\tau \int_0^{\sigma_1} d\sigma \sqrt{(\dot{X} \cdot X')^2 - (\dot{X})^2 (X')^2}.$$

The Nambu-Goto Action, named in honor of its inventors, the Japanese theorists Yoichiro Nambu and T. Goto, describes the "string action" of a one-dimensional string as it sweeps through a quantum-theoretical version of time. The variable "S" in the formula represents the total surface area of a string's worldsheet, that is, the area covered as it sweeps through spacetime. I share this bit of string theory esoterica not to intimidate you with highfalutin math, but to demonstrate that to decide to call the "object" within this formula a string is, in many respects, arbitrary. It is a matter of convenience that helps theorists to imagine more effectively the mathematics with which they work.

Sure, the image of the string may be more or less apt. Strings that we are familiar with in everyday experience behave in ways that are more or less appropriate for describing a theoretical object that might exist on the incredibly remote and alien Planck scale. But there is no abiding reason to assume that what we understand as objects on human scales—things we can see and grasp—have any meaning on scales where what we "see" and "grasp" are mediated by mathematical arguments and data collected from instruments.

In effect, when theorists "expose" string theory to themselves and to us, a lay audience, what we are getting is not the science *per se*, but a *scientific imaginary*. A scientific imaginary is a complex of images that gestures towards a coherent worldview. A worldview is a way of looking (a view) coupled to a world that is readily viewable. We imagine a new worldview by recombining in novel ways human-scale images from existing worldviews. These images are based on objects and events drawn from everyday experience—with which we are relatively familiar and thus, are relatively easy to comprehend. I stress the qualification "relatively," because it is the novelty in the recombinations of these images that makes a scientific imaginary feel weird and incredible. A scientific imaginary is most appealing when it manages to find that ideal mixture of strangeness and familiarity. It is just the right amount of weird.

A scientific imaginary offers us a whole world, even though it may not succeed in fulfilling its intention to be coherent. As a whole, a scientific imaginary engenders a worldview. The worldview is the whole—that amalgam of images and their rules of interaction that attempts to extrapolate the whole from its parts. As a reductionist strategy, string theory declares, in its crudest articulation, that the world is made of strings. In this sense, string theory is atomistic. Atomism here is synonymous with reductionism—a strategy whereby the whole is understood in terms of its smallest constituent parts. Reductionism is a kind of worldview where *the parts are more than the whole,* the reverse of holism.

But a worldview also, importantly, contains a view: not only is it a world, but also the human agency that engages with it. A worldview, in this sense, is synoptic. One view, formed through consensus, gathers together a whole. The two elements are stitched together—a world and those who would recognize it as such.

We can think of a string theory imaginary as a particular kind of worldview: as a cosmology. It is important here, though, to distinguish cosmology as an imaginary and cosmology as scientific practice. Within physics, cosmology is a distinct discipline. Cosmologists try to understand the universe on the largest of scales by means of theoretical models extrapolated from telescopic observation: the dynamics of solar systems, black holes, galaxies, and galaxy clusters. They work from the assumption that the universe on the vastest of scales can be understood as an ordered whole.

To speak of a string theory imaginary as a cosmology is to recognize that the word *cosmos* has a more expansive connotation than *world*. Where world often implies solely planet earth apprehended on human scales or global scales (for example, to speak of world peace), cosmos clearly designates the universe in all its vastness and totality. Yet the word cosmology also conjoins the outside with the inside, a *cosmos* with a *logos*, its apprehension and comprehension. For our purposes, then, a cosmology mediates the interaction between nature, the objective world out there in its fullest range from microscopic to macroscopic, and culture, the social agencies and practices of a human community. A cosmology, then, is an imaginary situated between culture and nature. It implies: a signifier, the cosmos; a signified, its cultural meaning; and those who are doing the signifying.

Contemplation of the cosmos has always held a great deal of emotional power. The word *cosmos* also comes from the ancient Greek. Originally, *cosmos* meant order or that which is well arranged. *The Oxford English Dictionary* defines *cosmos* as "the world or universe as an ordered and harmonious system." It was Pythagoras and his followers that extended the meaning of *cosmos* from "order" to the world because they saw in the world an "order and arrangement." For Pythagoras it was number that perfectly expressed the order in the world.

The word *universe* is the Roman counterpart to *cosmos*, and in Latin it literally means "one turn." The *OED* defines *universe* as "the whole of created or existing things regarded collectively; all things (including the earth, the heavens, and all the phenomena of space) considered as constituting a systematic whole." Perhaps one reason why we, more often than not, face the vastness of the cosmos with a sense of reverence and awe is that it defies direct apprehension. Even for the ancient Greeks, the world was a vast place with literal and figurative frontiers, past which there was only the unknown. In the face of this utter vastness, there's something incredibly bold about the act of defining the universe as one turn, one grand, sweeping act of imagination. How hubristic to conceive of the vast multitude of all things, ideas, and places—of which we are and will ever be only dimly aware—as one ordered and harmonic whole! There is also something incredibly wishful and innocent about it. Today, our image of the cosmos has changed significantly from that of the ancients. Those far-off frontiers that seemed so unassailable to them, we have painstakingly, over the course of centuries, pushed further and further back.

In her groundbreaking study of the particle physics community, *Beamtimes and Lifetimes*, anthropologist Sharon Traweek defines culture as "a group's shared set of meanings, its implicit and explicit messages, encoded in social action, about how to interpret experience." The converse of this definition also proves true: a culture is a shared set of social practices encoded in a symbolic structure. Action in the world determines what the world means, just as what that world means informs action.

Traweek's emphasis is sociological. She is referring to social action within the physics community: theorists doing calculations on scratch-pads, whiteboards, or computers; theorists attending conferences, conversing with other theorists, publishing papers, exchanging emails, etc. Unlike Traweek's work, the pieces in this collection are not concerned so much with showing the specific social practices of string theorists in the doing of physics and how those practices shape their understanding of the world.

Rather, in reading these pieces, notice how the writers portray, through their own particular version of a string theory imaginary, people interacting with each other *and* with things in the world. Within each string theory imaginary, then, we can examine the extent to which social action generates interpretation just as interpretation, that "shared set of meanings," generates social action. We can ask, to what extent are these two flows of meaning and interaction, in seemingly opposite directions, mutually constituting?

From this perspective, then, the contrast between nature and culture becomes more a matter of emphasis. Within an imaginary, the objective world becomes a projection of a community's self-regulating structure of social actions—and vice versa. This coupling of cosmos with culture, this cosmology, is an idea given currency by the early twentieth-century sociologist Émile Durkheim. Traweek paraphrases what she calls the "Durkheim supposition" thus:

> [A] culture's cosmology—its ideas about space and time and its explanation for the world—is reflected in the domain of social actions. In other words, ideas about time and space structure social relations, and the spatial and temporal patterns of human activity correspond to people's concepts of time and space.

A culture's notions of the world it inhabits (space and time) inform its social practices (patterns of activity) while its social practices shape its notions of the world.

In the context of a scientific imaginary such as string theory, then, to speak of culture is to focus on what the French philosopher Michel Serres calls "our relations among ourselves," while to evoke nature is to emphasize "our rapport with things." Serres uses the example of the post-war space program to highlight this interlacing of culture and nature: "Every technology transforms our rapport with things (the rocket takes off for the stratosphere) and, at the same time, our relations among ourselves (the rocket ensures publicity for the nations that launch it)." He continues: "This object, which we thought simply brought us into relationship with the stars, also brings us into relationship among ourselves." In this sense, scientific imaginaries cross-fertilize with "cultural" imaginaries, where an emphasis of orientation "out there" or "among us" determines an imaginary's status and function. Playing on the etymology of the words "rapport" and "relate," we "carry" how we imagine the cosmos back into the community and, in turn, we "carry" how we imagine the community back into the cosmos.

With string theory, the objects that concern us are not technological *per se*, like Sputnik or the Apollo Saturn V rocket, but theoretical. A string theory imaginary—and the things that populate it—finds its greatest currency in a consumer culture that puts a premium on the trafficking of scientific ideas, where the bandying about of scientific ideas becomes yet another means for displays of status. Scientific ideas become an intellectual surplus in an economy of exchange where the cachet of a given scientific idea stems from its novelty, its sexiness, its obscurity. A string theory imaginary finds its place in a culture of informationalism, a culture that marks scientific knowledge as the last frontier, where the unveiling of the hidden essence of nature becomes a peripatetic lurch at the virtual quasi-domestication of the alien on incredibly remote scales. It is an arguably myopic, if not solipsistic culture, where, as the sociologist Manuel Castells suggests, "[w]e are just entering a new stage in which

Culture refers to Culture, having superseded Nature to the point that Nature is artificially revived ('preserved') as a cultural form."

Keeping the "Durkheim supposition" in mind, let's explore four examples of scientific imaginaries. The first comes from Bertrand Russell's famous *The ABC of Relativity*, first published in 1925. In this passage, Russell draws a link between the physics concept of force and politics:

> If people were to learn to conceive the world in the new way, without the old notion of "force," it would alter not only their physical imagination, but probably also their morals and politics. [...] In the Newtonian theory of the solar system, the sun seems like a monarch whose behests the planets have to obey. In the Einsteinian world there is more individualism and less government than in the Newtonian.

This quote illustrates how a scientific imaginary works. Both the "Newtonian theory" and Einstein's special relativity are theories expressed in the language of mathematics. Russell conflates an expository explanation of these theories, what he calls "their physical imagination," with the theories themselves. This, in turn, allows for an easy imaginative leap to the discursive domains of "morals and politics." Sokal and Bricmont would certainly argue that such an imaginative leap betrays an inappropriate adaptation—a distortion. Russell hijacks the authority of Einstein's theory to champion a liberal (if not libertarian) political allegiance, one that valorizes individualism and decentralized government.

The second example comes from Mary Midgley in her book *Science and Poetry*: "[T]he social development of individualism increased the symbolic appeal of physical atomism, while the practical successes of physical atomism made social individualism look scientific." Like Russell, she writes of the relationship between Newtonian cosmology and "social individualism." In this cultural configuration, the image of the atom and the image of the human individual become mutually reinforcing. But rather than co-opting a scientific imaginary to promote a political view, Midgley calls attention to the coupling of that imaginary to the culture in which it flourishes. We get a sense that Midgley is aware of the fissure between actual Newtonian theory and a scientific imaginary based on it. Unlike a positivist perspective—that envisions a chasm between the conceptual content of scientific practice and the imaginative content of non-scientific discourse—Midgley here recognizes the amplifying feedback that a scientific imaginary can supply to an ideological agenda.

Scientific imaginaries fill the gaps between context-specific theoretical arguments so thoroughly that to conceive of the dichotomy between science and its ostensibly lesser sister, the imagination, as an unbridgeable rift is to deny the extent to which imaginaries support and sustain scientific arguments. To recognize this is to better understand the extent to which, in the case of a supposedly Newtonian cosmology, an ideology of "social individualism" can then appear scientific. Social individualists find it irresistible to co-opt science's authority to justify their ideological stance. In effect, a way of imagining the "out there" justifies the "among us."

The third example of a scientific imaginary is none other than Galileo's theory of heliocentrism. That the earth revolves around the sun would seem to be such a self-evident truth as to be unassailable. In the conventional account, heliocentrism represents a fundamental change in how we view the world, with all sorts of implications for how we should behave towards the world and towards each other. Yet such a worldview belies a subtle conflation akin to that of Russell, one that, on closer inspection, helps to further illustrate how scientific imaginaries work. In spite of our resolute conviction that the earth revolves around the sun, we nevertheless still have the daily, earth-bound experience of the sun revolving around the earth. From where we stand, the sun rises in the East and sets in the West. Obliquely (it's not good to stare for too long), we watch the sun slowly make its arcing journey across the sky.

What we moderns do now, though, *is imagine the earth moving around the sun, and mark that imaginary as the truth, the deeper reality*. That truth bears the authority of consensus over many generations; it is a highly stable knowledge. Most adherents of the heliocentric theory cannot prove its veracity: they accept it as a matter of dogma. We defer to the specialists who, if funded, would gladly conduct (and have conducted as a matter of public record) experiments to verify its truth. In the heliocentric imaginary, the relationship of earth to sun makes use of the geocentricity of common experience. It employs a structure of correspondences between the images (the sun and the planets) and astronomic observation (with the aid of telescopes), but reverses the dynamic and the scale (earth shrinks while sun expands). The heliocentric imaginary also draws upon reinforcing images of earth-images revolving around sun-images, including, but by no means limited to: declarative statements like "the earth revolves around the sun" given by authorities such as primary school science teachers; dioramas with fruit-sized painted styrofoam balls connected by rods and hinges; and graphic illustrations in textbooks or online video simulations.

Galileo used his telescope to observe the Milky Way. He did not recognize it as a galaxy, since at the time there was no basis for comparison between it as an object and similar objects. What Galileo found beyond the Milky Way were fuzzy objects that were clearly not stars. These objects came to be called nebulae. Only later—much later, in the 1920s—with the aid of more powerful telescopes, did Edwin Hubble understand that many of those fuzzy nebulae were actually galaxies. It was only then that galaxies (from the Greek, meaning "milk") became a definitive feature of the known universe. Before then, we lived, with respect to our scientific imaginary, in a world without galaxies.

In our collective cosmology, one imaginary does not merely replace the other. We hold in our mind's eye the daily geocentric experience *and* the heliocentric imaginary, which contradicts while also "clarifying" that experience. Our world has thus become all the more complex and collective, for now we depend on ever more specialists to reveal to us and each other, by means of imaginaries, the various "deeper" realities. This world becomes populated with ever more intricate networks of sometimes complementary, sometimes contradictory imaginaries. Mind you, I am not suggesting that the earth does not, in fact, revolve around the sun—that heliocentrism is just in our head. Rather, our access to the world "out there" is irrevocably mediated by imaginaries. There can be no direct and self-evident apprehension of the world. As a consequence, we perpetually rely on the authority of others to understand the world. And we regulate our interactions with that world accordingly. The myriad and disparate scientific methods may not be faith-based, but belief in one monologic scientific worldview surely is.

As our fourth example, let's return to quantum theory and its effect on the image of the atom. In the late nineteenth and early twentieth centuries, popular culture became fascinated with this startling new science. The atom had previously been understood to embody its etymology—from the Greek, meaning "indivisible" or more literally, "uncuttable." Quantum theory shows that the atom is, in fact, divisible. It consists of a nucleus and an arrangement of particles, electrons, that orbit this nucleus. Based on calculations and collider experiments, the scale of the atom as a whole (10^{-11} meters) and its nucleus (10^{-15} meters) suggest that there exists a vast amount of "empty space" within which electrons "orbit."

At the time, this new way of imagining atoms was understood as the deeper reality. That image shows the atom to be a field of empty space populated with various sparsely situated objects: a tiny nucleus with electrons rushing around it. This image superseded the prevailing image of the atom as a solid, whole, indivisible object. An authoritative knowledge (from physicists) privileged the former imaginary over the latter, often mustering for its reinforcement a higher-order image of surface and depth, where the "deeper" image is the one that is "true." As we imagine ourselves "going into" an atom, the empty space inside it becomes readily apparent.

Again, I am not suggesting that quantum theory is not true—a figment of the imagination. What I am saying is that each imaginary—whether an atom as indivisible object or an atom as field of empty space populated by various objects—is appropriate for a given context. Both describe one aspect of the reality of the world "out there." The actual methods of quantum theory—its mathematics and experiment design—provide prompts for intervention by our bodies with the world. Imaginaries reflect and reinforce those prompts. In a tangible sense, we grope our way blindly through the world with imaginaries serving as visual templates for that groping. Scientific imaginaries are, in effect, *productive* collective hallucinations.

In this collection Joseph Radke has a poem, "Life x 10^{-33}," that expresses this conceit elegantly:

> No, we can't renounce
> the invisible, the fluid foundation
> of the solidly seen. We can
> only imagine and speak
> in shrinking untruths.

Along those lines, consider this: heliocentrism may well be easy enough for a lay person to validate experimentally with the proper guidance and some perspicacity, but what about the existence of quarks—or strings, for that matter?

This collection explores the interplay between culture and nature, between how a community acts in the world and how its members communicate with and come to know each other. It asks us to re-imagine, to quote Serres once again, "our relations among ourselves" in light of "our rapport with things," and in particular, the wondrous new things of string theory. Each piece asks, in its own unique way: as social creatures through and through, just how do we project out into the cosmos our everyday concerns—and how does the cosmos reflect back on those concerns?

Accordingly, perhaps our universe is much more than just "one turn." Perhaps any notion of the universe necessarily involves an intricate folding over and doubling back—a feedback loop. Just as our imagining of the universe influences how we go about our business in this sticky matter of living in the human-scale world, the way we live in community influences how we imagine the universe.

In his Foreword to Lawrence Krauss's *The Physics of Star Trek*, Stephen Hawking observes that "there is a two-way trade between science fiction and science. Science fiction suggests ideas that scientists incorporate into their theories, but sometimes science turns up notions that are stranger than any science fiction." Hawking believes that, apart from being "good fun," science fiction "serves a serious purpose, that of expanding the human imagination."

The "atom smashers" played a pivotal role in putting an end to World War II by helping to invent the atomic bomb. Their success brought with it not only a great deal of prestige, but also political influence, and with that, a huge infusion of cash to the research universities where they worked after the war. In the romance of theoretical physics, the "atom smashers" became a vanguard exploring the frontiers of the cosmos, on both the tiniest and the vastest of scales. As such, physicists have become the *de facto* guardians and spokespersons of the primordial emotional appeal of cosmology.

Nevertheless, Sokal and Bricmont, as purists of the "hard sciences," are quick to point out: "scientific theories are not like novels; in a scientific context [...] words have specific meanings, which differ in subtle but crucial ways from their everyday meanings, and which can only be understood within a complex web of theory and experiment." Of course, they warn that scientists are always not entirely innocent in this abuse of science. They may inadvertently encourage "fashionable nonsense":

> [W]ell-known scientists, in their popular writings, often put forward speculative ideas as if they were well-established, or extrapolate their results far beyond the domain where they have been verified. Finally, there is a damaging tendency—exacerbated, no doubt, by the demands of marketing—to see a "radical conceptual revolution" in each innovation. All these factors combined give the educated public a distorted view of scientific activity.

But what can we say about the tempting delectability of string theory imaginaries for we whose business it is to "expand the human imagination"?

I give you the Austrian philosopher Paul Feyerabend's rebuttal of scientific realism:

> Knowledge so conceived is not a series of self-consistent theories that converges towards an ideal view; it is not a gradual approach to truth. It is rather an ever increasing *ocean of mutually incompatible alternatives*, each single theory, each fairy-tale, each myth that is part of the collection forcing the others into greater articulation and all of them contributing, via this process of competition, to the development of our consciousness.

Perhaps realist hardliners ought not to be so automatic in admonishing those who would "distort" science to tell stories about people or to grapple with that great question of our place in the world. My doctoral research focuses on the scientific imaginaries that constitute string theory. The remarkable thing to me is that once you start to look for them, you find these imaginaries popping up in not so obvious places. They're there in the technical discourse, however muted or opaque, just as much as they're there in a poem or a sci-fi short story like the one in this collection by Adam Roberts called "S-Bomb." The story explores, playing off string imagery in fascinating ways, the unraveling of the social fabric.

Perhaps we shouldn't be so quick to dismiss literary treatments of string theory as mere flights of fancy, utterly unmoored from its mathematical truth. Rather let us approach string theory imaginaries as native to many habitats and as such, adapted to those varying habitats. As writers and readers, we can take up strings and branes in our hands, knowing full well that such things come from a place to which they are intricately and precisely fitted, and—with a knowing wink—redeploy these wondrous cosmic objects to places that at first inspection might seem like ill-suited homes. This will nurture within us a more nuanced understanding of the social world in terms of the cosmic, and in turn, a finer understanding of the cosmic in terms of the social.

Colette Inez's poem, "Cosmic Gambol," epitomizes, with a knowing wink, the coupling of a cosmos with an *us*:

> It's a narrow path
> to find my boson mate, elusive Higgs,
> as I desire my *colettions*
> to loop and loop in wavicles
> of joy that make me matter.

Is this coupling not unavoidable, burdened, as we are, with all that relating, so often haphazard, to each other and to things? How pure can science really be? Is it not always laden with feeling—however much hard science purists would try to convince themselves and those in awe of their work that what they offer us is utterly purified of any contaminating human subjectivity?

I'm not saying that the world out there is merely a social construct—a fantasy that we delude ourselves into believing is real. I'm saying that as human beings, we are creatures that filter, and that what is available to us in the form of sensory contact with the universe is severely curtailed by the limits of those senses, which after all, evolved to suit an engagement with the world on human body scales that would best ensure our survival. As I suggested earlier, in some sense, we are blind to the cosmos in all its awesome totality. The sad and humbling fact is that we are only dimly aware of a small fraction of all its glory. (How different the world must appear to a dog, or an ant, or a bacterium.)

I must say that I'm amazed by the startling range of form and expression in this collection—and I hope you will be too. To borrow an expression from Douglas Hofstadter, this collection engenders a kind of "strange loop." It is string theory craning its neck and, turning around, gazing back upon its own form—in all its multiple facets. (Perhaps it is more of a strange hairball or knot—or a many-headed hydra contemplating its own fertile plurality.) Of course, such a collection cannot be exhaustive nor can it pretend to be the final word, only an injection of fresh perspectives from vantage points not necessarily native to the exclusive coterie of working physicists (though, a few of the essays are).

To reiterate, the question on which *Riffing on Strings* pivots is: what can we say and feel about the cosmos we live in through the prism of string theory imaginaries? Rather than just passively consuming science, in accepting its supposed monologic authority, the writers who've contributed to this collection revel in the pluralities of string theory. To quote Feyerabend again:

> [A] uniform "scientific view of the world" may be useful *for people doing science*—it gives them motivation without tying them down. It is like a flag. Though presenting a single pattern it makes people do many different things. However, *it is a disaster for outsiders* (philosophers, fly-by-night mystics, prophets of the New Age). It suggests to them the most narrowminded religious commitment and encourages a similar narrowmindedness on their part.

A string theory imaginary lends itself to analogies with fabric, weaving, unraveling—and to those of music, resonance, and harmony. The writers in this collection implicitly ask, how may we riff on strings? That is, how may we explore the imaginaries inherent in string theory in ways that offer new perspectives on the world we live in, both the social world—the world within our grasp and within earshot—and the vast and enveloping cosmos of which we are a small but significant (at least to us) part? As in music, a meticulous assonance is not always the most moving—often, we long for some dissonance and distortion to complement and highlight our sense of harmony.

I hope that having been inspired by string theory, the pieces in this collection will, in turn, inspire you, the reader, to look upon your world anew. After all, are we not all—at least in some small part—philosophers, fly-by-night mystics, prophets of the New Age? As Jeff P. Jones suggests in his poem in this collection, "Raise It Up in the Mind of Me," a string theory imaginary allows us to "inhabit more space than [we] ought."

I would love to comment here—with all due appreciation—on every piece contained in *Riffing on Strings*. But given the prolixity of this, my introductory disquisition, I've no doubt already tried your patience to its limits. I can only hope I've offered you a few useful keys for decoding the cosmic secrets contained herein.

To be honest, we agonized over the best way to organize the collection. It felt a bit like one of those standardized exam questions where you're asked to "Select the one that is different from the others." You've been told emphatically there's a correct answer, which should be obvious. (Take, for example, this classic IQ test question: Which one of these is least like the other four? A) Horse B) Kangaroo C) Goat D) Deer E) Donkey. Can you think of a good reason to exclude each of them?) Likewise, as we pondered our stack of pieces, we found perfectly valid reasons for sorting them in several different ways. We considered sorting them by: tone, style, thematic affinity, random shuffling, on our own imperfect sense of what we like best, name recognition where writers with bigger reputations go first, or, with a flinging up of the hands, simply editors' caprice. In the end, we decided to organize the collection based on the more prosaic criterion of genre. We start with essays, follow them with short stories, then poetry, and then finish with a complete version of Carole Buggé's remarkable play, *Strings*. We respectfully leave it to the reader to trace her own thread of continuity through this labyrinthine arrangement. By all means, feel free to start at the beginning, turn one page after the other, and finish at the end.

Or, if the spirit moves you, we also encourage you to careen serendipitously from one piece to another…

Notes

xiv "To get a sense of scale…" Brian Greene, *The Elegant Universe: Superstrings, Hidden Dimensions, and the Quest for the Ultimate Theory* (New York: Random House, 1999) 130.
xv "…the so-called Two Cultures…" C.P. Snow, *The Two Cultures and The Scientific Revolution* (Cambridge: Cambridge UP, 1959).
xvi "In quantum gravity…" Alan Sokal, "Transgressing the Boundaries: Towards a Transformative Hermeneutics of Quantum Gravity," *Social Text* 46/47 (1996): 217-252, 25 Jan 08 < http://www.physics.nyu.edu/~as2/transgress_v2/transgress_v2_singlefile.html >.
xvi "When concepts from mathematics or physics…" Alan Sokal and Jean Bricmont, *Fashionable Nonsense: Postmodern Intellectuals' Abuse of Science* (New York: Picador, 1998) 9.
xix "I'm working on this theory…" Leonard Susskind, *The Cosmic Landscape: String Theory and the Illusion of Intelligent Design* (New York: Little, Brown, 2005) 219.
xxiv "…we should observe a universe…" P.C.W. Davies, "Inflation in the Universe and Time Asymmetry," *Nature* Vol 312:5994 (6 December 1984): 525.
xxvii "Even now, one can go to a conference…" Lee Smolin, *Three Roads to Quantum Gravity* (New York: Basic Books, 2001) 181.
xxviii "I realized how much attention…" Lisa Randall, *Warped Passages: Unraveling the Mysteries of the Universe's Hidden Dimensions* (New York: Ecco, 2005) 315.
xxix "The Literal Meaning Theory entails another position…" George Lakoff and Mark Johnson, *More Than Cool Reason: A Field Guide to Poetic Metaphor* (Chicago : U of Chicago Press, 1989) 123.
xxxii "…a group's shared set of meanings…" Sharon Traweek, *Beamtimes and Lifetimes: The World of High Energy Physicists* (Cambridge, MA: Harvard UP, 1988) 7.
xxxii "…what she calls the 'Durkheim supposition' thus…" Traweek 157.
xxxiii "…our relations among ourselves…" Michel Serres and Bruno Latour, *Conversations on Science, Culture, and Time*, trans. Roxanne Lapidus (East Lansing, MI: U. of Michigan Press, 1995) 141.
xxxiii "This object, which we thought…" Serres 148.
xxxiii "…[w]e are just entering a new stage…" Manuel Castells, *The Rise of the Network Society* (Oxford: Blackwell, 1996) 477.
xxxiv "If people were to learn to conceive the world…" Bertrand Russell, *The ABC of Relativity*, 5th rev. ed. (London: Routledge, 1997) 2.
xxxiv "[T]he social development of individualism…" Mary Midgley, *Science and Poetry* (New York: Routledge, 2001) 10.
xxxviii "…there is a two-way trade between science fiction and science…" Stephen Hawking, "Foreward," *The Physics of Star Trek*, by Lawrence Krauss (New York: Basic, 1995) xii, xi.
xxxviii "…scientific theories are not like novels…" Sokal and Bricmont 187.
xxxviii "[W]ell-known scientists, in their popular writings…" Sokal and Bricmont 193.

xxxix "Knowledge so conceived..." Paul Feyerabend, *Against Method* (London: Verso, 1993) 21.
xl "[A] uniform 'scientific view of the world' may be useful..." Feyerabend 250.

Essays

"Superstring" by Félix Sorondo

M-Theory: The Mother of All Superstrings

Michio Kaku

Every decade or so, a stunning breakthrough in string theory sends shock waves racing through the theoretical physics community, generating a feverish outpouring of papers and activity. In 1995, the Internet lines were burning up as papers kept pouring into the Los Alamos National Laboratory's computer bulletin board, the official clearing house for superstring papers. John Schwarz of Caltech, for example, spoke to conferences around the world proclaiming the "second superstring revolution." Edward Witten of the Institute for Advanced Study in Princeton gave a spell-binding 3 hour lecture describing it. The aftershocks of the breakthrough even have been shaking other disciplines, like mathematics. The director of the Institute, mathematician Phillip Griffiths, said, "The excitement I sense in the people in the field and the spin-offs into my own field of mathematics…have really been quite extraordinary. I feel I've been very privileged to witness this first hand."

Cumrun Vafa at Harvard has said, "I may be biased on this one, but I think it is perhaps the most important development not only in string theory, but also in theoretical physics at least in the past two decades." What triggered all this excitement was the discovery of something called "M-theory," a theory which may explain the origin of strings. In one dazzling stroke, this new M-theory has solved a series of long-standing puzzling mysteries about string theory which have dogged it from the beginning, leaving many theoretical physicists (myself included!) gasping for breath. M-theory, moreover, may even force string theory to change its name. Although many features of M-theory are still unknown, it does not seem to be a theory purely of strings. Michael Duff of Texas A & M has already been giving speeches with the title "The theory formerly known as strings!" String theorists are careful to point out that this does not prove the final correctness of the theory. Not by any means. That may take years or decades more. But it marks a most significant breakthrough that is already reshaping the entire field.

Parable of the Lion

Einstein once said, "Nature shows us only the tail of the lion. But I do not doubt that the lion belongs to it even though he cannot at once reveal himself because of his enormous size." Einstein spent the last 30 years of his life searching for the "tail" that would lead him to the "lion," the fabled unified field theory or the "theory of everything," which would unite all the forces of the universe into a single equation. The four forces (gravity, electromagnetism, and the strong and weak nuclear forces) would be unified by an equation perhaps one inch long. Capturing the "lion" would be the greatest scientific achievement in all of physics, the crowning achievement of 2,000 years of scientific investigation, ever since the Greeks first asked themselves what the world was made of. But although Einstein was the first one to set off on this noble hunt and track the footprints left by the lion, he ultimately lost the trail and wandered off into the wilderness. Other giants of 20th century physics, like Werner Heisenberg and Wolfgang Pauli, also joined in the hunt. But all the easy ideas were tried and shown to be wrong. When Niels Bohr once heard a lecture by Pauli explaining his version of the unified field theory, Bohr stood up and said, "We in the back are all agreed that your theory is crazy. But what divides us is whether your theory is crazy enough!"

The trail leading to the unified field theory, in fact, is littered with the wreckage of failed expeditions and dreams. Today, however, physicists are following a different trail which might be "crazy enough" to lead to the lion. This new trail leads to superstring theory, which is the best (and in fact only) candidate for a theory of everything. Unlike its rivals, it has survived every blistering mathematical challenge ever hurled at it. Not surprisingly, the theory is a radical, "crazy" departure from the past, being based on tiny strings vibrating in 10 dimensional space-time. Moreover, the theory easily swallows up Einstein's theory of gravity. Witten has said, "Unlike conventional quantum field theory, string theory requires gravity. I regard this fact as one of the greatest insights in science ever made." But until recently, there has been a glaring weak spot: string theorists have been unable to probe all solutions of the model, failing miserably to examine what is called the "non-perturbative region," which I will describe shortly. This is vitally important, since ultimately our universe (with its wonderfully diverse collection of galaxies, stars, planets, subatomic particles, and even people) may lie in this "non-perturbative region." Until this region can be probed, we don't know if string theory is a theory of everything—or a theory of nothing! That's what today's excitement is all about. For the first time, using a powerful tool called "duality," physicists are now probing beyond just the tail, and finally seeing the outlines of a huge, unexpectedly beautiful lion at the other end. Not knowing what to call it, Witten has dubbed it "M-theory." In one stroke, M-theory has solved many of the embarrassing features of the theory, such as why we have 5 superstring theories. Ultimately, it may solve the nagging question of where strings come from.

"Pea Brains" and the Mother of all Strings

Einstein once asked himself if God had any choice in making the universe. Perhaps not, so it was embarrassing for string theorists to have five different self-consistent strings, all of which can unite the two fundamental theories in physics, the theory of gravity and the quantum theory.

Each of these string theories looks completely different from the others. They are based on different symmetries, with exotic names like E(8)xE(8) and O(32).

Not only this, but superstrings are in some sense not unique: there are other non-string theories which contain "supersymmetry," the key mathematical symmetry underlying superstrings. (Changing light into electrons and then into gravity is one of the rather astonishing tricks performed by supersymmetry, which is the symmetry that can exchange particles with half-integral spin, like electrons and quarks, with particles of integral spin, like photons, gravitons, and W-particles.

In 11 dimensions, in fact, there are alternate super theories based on membranes as well as point particles (called supergravity). In lower dimensions, there is moreover a whole zoo of super theories based on membranes in different dimensions. (For example, point particles are 0-branes, strings are 1-branes, membranes are 2-branes, and so on.) For the *p*-dimensional case, some wag dubbed them *p*-branes (pronounced "pea brains"). But because *p*-branes are horribly difficult to work with, they were long considered just a historical curiosity, a trail that led to a dead-end. (Michael Duff, in fact, has collected a whole list of unflattering comments made by referees to his National Science Foundation grant concerning his work on *p*-branes. One of the more charitable comments from a referee was: "He has a skewed view of the relative importance of various concepts in modern theoretical physics.") So that was the mystery. Why should supersymmetry allow for 5 superstrings and this peculiar, motley collection of *p*-branes? Now we realize that strings, supergravity, and *p*-branes are just different aspects of the same theory. M-theory (M for "membrane" or the "mother of all strings," take your pick) unites the 5 superstrings into one theory and includes the *p*-branes as well. To see how this all fits together, let us update the famous parable of the blind wise men and the elephant. Think of the blind men on the trail of the lion. Hearing it race by, they chase after it and desperately grab onto its tail (a one-brane). Hanging onto the tail for dear life, they feel its one-dimensional form and loudly proclaim "It's a string! It's a string!"

But then one blind man goes beyond the tail and grabs onto the ear of the lion. Feeling a two-dimensional surface (a membrane), the blind man proclaims, "No, it's really a two-brane!" Then another blind man is able to grab onto the leg of the lion. Sensing a three-dimensional solid, he shouts, "No, you're both wrong. It's really a three-brane!" Actually, they are all right. Just as the tail, ear, and leg are different parts of the same lion, the string and various *p*-branes appear to be different limits of the same theory: M-theory. Paul Townsend of Cambridge University, one of the architects of this idea, calls it "*p*-brane democracy," i.e. all p-branes (including strings) are created equal. Schwarz puts a slightly different spin on this. He says, "we are in an Orwellian situation: all *p*-branes are equal, but some (namely strings) are more equal than others. The point is that they are the only ones on which we can base a perturbation theory." To understand unfamiliar concepts such as duality, perturbation theory, non-perturbative solutions, it is instructive to see where these concepts first entered into physics.

Duality

The key tool to understanding this breakthrough is something called "duality." Loosely speaking, two theories are "dual" to each other if they can be shown to be equivalent under a certain interchange. The simplest example of duality is reversing the role of electricity and magnetism in the equations discovered by James Clerk Maxwell of Cambridge University 140 years ago. These are the equations which govern light, TV, X-rays, radar, dynamos, motors, transformers, even the Internet and computers. The remarkable feature about these equations is that they remain the same if we interchange the magnetic B and electric E fields and also switch the electric charge e with the magnetic charge g of a magnetic "monopole": E <--> B and e <--> g (In fact, the product eg is a constant.) This has important implications. Often, when a theory cannot be solved exactly, we use an approximation scheme. In first year calculus, for example, we recall that we can approximate certain functions by Taylor's expansion. Similarly, since e^2 = 1/137 in certain units and is hence a small number, we can always approximate the theory by power expanding in e^2. So we add contributions of order e^2 + e^4 + e^6 etc. in solving for, say, the collision of two particles. Notice that each contribution is getting smaller and smaller, so we can in principle add them all up. This generalization of Taylor's expansion is called "perturbation theory," where we perturb the system with terms containing e^2. For example, in archery, perturbation theory is how we aim our arrows. With every motion of our arms, our bow gets closer and closer to aligning with the bull's eye.) But now try expanding in g^2. This is much tougher; in fact, if we expand in g^2, which is large, then the sum $g^2 + g^4 + g^6$ etc. blows up and becomes meaningless. This is the reason why the "non-perturbative" region is so difficult to probe, since the theory simply blows up if we try to naively use perturbation theory for large coupling constant g. So at first it appears hopeless that we could ever penetrate into the non-perturbative region. (For example, if every motion of our arms got bigger and bigger, we would never be able to zero in and hit the target with the arrow.) But notice that because of duality, a theory of small e (which is easily solved) is identical to a theory of large g (which is difficult to solve). But since they are the same theory, we can use duality to solve for the non-perturbative region.

S, T, and U Duality

The first inkling that duality might apply in string theory was discovered by K. Kikkawa and M. Yamasaki of Osaka Univ. in 1984. They showed that if you "curled up" one of the extra dimensions into a circle with radius R, the theory was the same if we curled up this dimension with radius $1/R$. This is now called T-duality: R <--> $1/R$. When applied to various superstrings, one could reduce 5 of the string theories down to 3 (see figure). In 9 dimensions (with one dimension curled up) the Type IIa and IIb strings are identical, as are the E(8)xE(8) and O(32) strings.

Unfortunately, T-duality is still a perturbative duality. The next breakthrough came when it was shown that there was a second class of dualities, called S-duality, which provides a duality between the perturbative and non-perturbative regions of string theory. Another duality, called U-duality, is even more powerful.

Then Nathan Seiberg and Witten brilliantly showed how another form of duality could solve for the non-perturbative region in four dimensional supersymmetric theories. However, what finally convinced many physicists of the power of this technique was the work of Paul Townsend and Edward Witten. They caught everyone by surprise by showing that there is a duality between 10 dimensional Type IIa strings and 11 dimensional supergravity! The non-perturbative region of Type IIa strings, which was previously a forbidden region, was revealed to be governed by 11 dimensional supergravity theory, with one dimension curled up. At this point, I remember that many physicists (myself included) were rubbing our eyes, not believing what we were seeing. I remember saying to myself, "But that's impossible!"

All of a sudden, we realized that perhaps the real "home" of string theory was not 10 dimensions, but possibly 11, and that the theory wasn't fundamentally a string theory at all! This revived tremendous interest in 11 dimensional theories and *p*-branes. Lurking in the 11th dimension was an entirely new theory which could reduce down to 11 dimensional supergravity as well as 10 dimensional string theory and *p*-brane theory.

Detractors of String Theories

To the critics, however, these mathematical developments still don't answer the nagging question: how do you test it? Since string theory is really a theory of Creation, when all its beautiful symmetries were in their full glory, the only way to test it, the critics wail, is to recreate the Big Bang itself, which is impossible. Nobel Laureate Sheldon Glashow likes to ridicule superstring theory by comparing it with former President Reagan's *Star Wars* plan, i.e. they are both untestable, soak up resources, and both siphon off the best scientific brains.

Actually, most string theorists think these criticisms are silly. They believe that the critics have missed the point. The key point is this: if the theory can be solved non-perturbatively using pure mathematics, then it should reduce down at low energies to a theory of ordinary protons, electrons, atoms, and molecules, for which there is ample experimental data. If we could completely solve the theory, we should be able to extract its low energy spectrum, which should match the familiar particles we see today in the Standard Model. Thus, the problem is not building atom smashers 1,000 light years in diameter; the real problem is raw brain power. If only we were clever enough, we could write down M-theory, solve it, and settle everything.

Evolving Backwards

So what would it take to actually solve the theory once and for all and end all the speculation and back-biting? There are several approaches. The first is the most direct: try to derive the Standard Model of particle interactions, with its bizarre collection of quarks, gluons, electrons, neutrinos, Higgs bosons, etc. etc. etc. (I must admit that although the Standard Model is the most successful physical theory ever proposed, it is also one of the ugliest.) This might be done by curling up 6 of the 10 dimensions, leaving us with a 4 dimensional theory that might resemble the Standard Model a bit. Then try to use duality and M-theory to probe its non-perturbative region, seeing if the symmetries break in the correct fashion, giving us the correct masses of the quarks and other particles in the Standard Model. Witten's philosophy, however, is a bit different. He feels that the key to solving string theory is to understand the underlying principle behind the theory.

Let me explain. Einstein's theory of general relativity, for example, started from first principles. Einstein had the "happiest thought in his life" when he leaned back in his chair at the Bern patent office and realized that a person in a falling elevator would feel no gravity. Although physicists since Galileo knew this, Einstein was able to extract from this the Equivalence Principle. This deceptively simple statement (that the laws of physics are indistinguishable locally in an accelerating or a gravitating frame) led Einstein to introduce a new symmetry to physics—general co-ordinate transformations. This in turn gave birth to the action principle behind general relativity, the most beautiful and compelling theory of gravity. Only now are we trying to quantize the theory to make it compatible with the other forces. So the evolution of this theory can be summarized as: Principle -> Symmetry -> Action -> Quantum Theory. According to Witten, we need to discover the analog of the Equivalence Principle for string theory. The fundamental problem has been that string theory has been evolving "backwards." As Witten says, "string theory is 21st century physics which fell into the 20th century by accident." We were never "meant" to see this theory until further on this century.

Is the End in Sight?

Vafa recently added a strange twist to this when he introduced yet another mega-theory, this time a 12 dimensional theory called F-theory (F for "father") which explains the self-duality of the IIb string. (Unfortunately, this 12 dimensional theory is rather strange: it has two time co-ordinates, not one, and actually violates 12 dimensional relativity. Imagine trying to live in a world with two times! It would put an episode of Twilight Zone to shame.) So is the final theory 10, 11, or 12 dimensional?

Schwarz, for one, feels that the final version of M-theory may not even have any fixed dimension. He feels that the true theory may be independent of any dimensionality of space-time, and that 11 dimensions only emerge once one tries to solve it. Townsend seems to agree, saying "the whole notion of dimensionality is an approximate one that only emerges in some semi-classical context." So does this means that the end is in sight, that we will someday soon derive the Standard Model from first principles? I asked some of the leaders in this field to respond to this question. Although they are all enthusiastic supporters of this revolution, they are still cautious about predicting the future. Townsend believes that we are in a stage similar to the old quantum era of the Bohr atom, just before the full elucidation of quantum mechanics. He says, "We have some fruitful pictures and some rules analogous to the Bohr-Sommerfeld quantization rules, but it's also clear that we don't have a complete theory."

Duff says, "Is M-theory merely a theory of supermembranes and super 5-branes requiring some (as yet unknown) non-perturbative quantization, or (as Witten believes) are the underlying degrees of freedom of M-theory yet to be discovered? I am personally agnostic on this point." Witten certainly believes we are on the right track, but we need a few more "revolutions" to finally solve the theory. "I think there are still a couple more superstring revolutions in our future, at least. If we can manage one more superstring revolution a decade, I think that we will do all right," he says. Vafa says, "I hope this is the 'light at the end of the tunnel' but who knows how long the tunnel is!" Schwarz, moreover, has written about M-theory: "Whether it is based on something geometrical (like supermembranes) or something completely different is still not known. In any case, finding it would be a landmark in human intellectual history." Personally, I am optimistic. For the first time, we can see the outline of the lion, and it is magnificent. One day, we will hear it roar.

Desperately Seeking Superstrings?

Paul Ginsparg and Sheldon Glashow

Why is the smart money all tied up in strings? Why is so much theoretical capital expended upon the properties of supersymmetric systems of quantum strings propagating in ten-dimensional space-time? The good news is that superstring theory may have the right stuff to explain the "low-energy phenomena" of high-energy physics and gravity as well. In the context of possible quantum theories of gravity, each of the few currently known superstring theories may even be unique, finite, and self-consistent. In principle a superstring theory ordains what particles exist and what properties they have, using no arbitrary or adjustable parameters. The bad news is that years of intense effort by dozens of the best and the brightest have yielded not one verifiable prediction, nor should any soon be expected. Called "the new physics" by its promoters, it is not even known to encompass the old and established Standard Model.

In lieu of the traditional confrontation between theory and experiment, superstring theorists pursue an inner harmony where elegance, uniqueness, and beauty define truth. The theory depends for its existence upon magical coincidences, miraculous cancellations, and relations among seemingly unrelated (and possibly undiscovered) fields of mathematics. Are these properties reasons to accept the reality of superstrings? Do mathematics and aesthetics supplant and transcend mere experiment? Will the mundane phenomenological problems that we know as physics simply come out in the wash in some distant tomorrow? Is further experimental endeavor not only difficult and expensive but unnecessary and irrelevant? Contemplation of superstrings may evolve into an activity as remote from conventional particle physics as particle physics is from chemistry, to be conducted at schools of divinity by future equivalents of medieval theologians. For the first time since the Dark Ages, we can see how our noble search may end, with faith replacing science once again. Superstring sentiments eerily recall "arguments from design" for the existence of a supreme being. Was it only in jest that a leading string theorist suggested that "superstrings may prove as successful as God, Who has after all lasted for millennia and is still invoked in some quarters as a Theory of Nature"?

The trouble began with quantum chromodynamics, an integral part of the Standard Model that underlies the quark structure of nucleons and the nuclear force itself. QCD is not merely a theory but, within a certain context, the theory of the strong force: It offers a complete description of nuclear and particle physics at accessible energies. While most questions are computationally too difficult for QCD to answer fully, it has had many qualitative (and a few quantitative) confirmations. That QCD is almost certainly "correct" suggests and affirms the belief that elegance and uniqueness—in this case, reinforced by experiment—are criteria for truth.

No observed phenomenon disagrees with or demands structure beyond the Standard Model. No internal contradictions and few loose ends remain, but there are some vexing puzzles: Why is the gauge group what it is, and what provides the mechanism for its breakdown? Why are there three families of fundamental fermions when one would seem to suffice? Aren't 17 basic particles and 17 tunable parameters too many? What about a quantum theory of gravity? Quantum field theory doesn't address these questions, and one can understand its greatest past triumphs without necessarily regarding it as fundamental. Field theory is clearly not the end of the story, so something smaller and better is needed: enter the superstring.

The trouble is that most of superstring physics lies up at the Planck mass—about 10^{19} GeV—and it is a long and treacherous road down to where we can see the light of day. A naive comparison of length scales suggests that to calculate the electron mass from superstrings would be a trillion times more difficult than to explain human behavior in terms of atomic physics. Superstring theory, unless it allows an approximation scheme for yielding useful and testable physical information, might be the sort of thing that Wolfgang Pauli would have said is "not even wrong." It would continue to attract newcomers to the field simply because it is the only obvious alternative to explaining why certain detectors light up like video games near the end of every funding cycle.

In the old days we moved up in energy step by step, seeing smaller and smaller structures. Observations led to theories or models that suggested further experiments. The going is getting rougher; colliders are inordinately expensive, detectors have grown immense, and interesting collisions are rare. Not even a politically popular "Superstring Detection Initiative" with a catchy name like "String Wars" could get us to energies where superstrings are relevant. We are stuck with a gap of 16 orders of magnitude between theoretical strings and observable particles, unbridgeable by any currently envisioned experiment. Conventional grand unified theories, which also depend on a remote fundamental energy scale (albeit one extrapolated upward from known phenomena rather than downward from abstract principle), retain the grand virtue that, at least in their simplest form, they were predictive enough to be excluded—by our failure to observe proton decay.

How tempting is the top-down approach! How satisfying and economical to explain everything in one bold stroke of our aesthetic, mathematical or intuitive sensibilities, thus displaying the power of positive thinking without requiring tedious experimentation! But a priori arguments have deluded us from ancient Greece on. Without benefit of the experimental provocation that led to Maxwell's equations and, inevitably, to the special theory of relativity, great philosophers pondering for millennia failed even to suspect the basic kinematical structure of space-time. Pure thought could not anticipate the quantum.

And even had Albert Einstein succeeded in the quest that consumed the latter half of his life, somehow finding a framework for unifying electromagnetism and gravity, we would by now have discarded his theory in the light of experimental data to which he had no access. He had to fail, simply because he didn't know enough physics. Today we can't exclude the possibility that micro-unicorns might be thriving at a length scale of 10^{-18} cm. Einstein's path, the search for unification now, is likely to remain fruitless.

Having a potentially plausible candidate "theory of everything" does dramatically alter the situation. But we who are haunted by the lingering suspicion that superstrings, despite all the hoopla, may be correct are likely to remain haunted for the foreseeable future. Only a continued influx of experimental ideas and data can allow the paths from top and bottom to meet. The theory of everything may come in its time, but not until we are certain that Nature has exhausted her bag of performable tricks.

Selling Science: String Theory and "Science Porn"

Kristine Larsen

As a Professor of Physics and Astronomy, it often feels as if I could retire today if every member of the general public who had asked me my thoughts on string theory prefaced their usually enthusiastic query with a dollar bill. From mail carriers to English professors, bookstore salespersons to hotel desk clerks, whenever I mention my career path to someone outside of the scientific sphere, I am undoubtedly asked about the dreaded "S-word." Why? How did a highly mathematical (and some have claimed largely metaphysical) technique of theoretical physics become so firmly ingrained in the public psyche? More importantly, why have more experimentally-relevant (and in the minds of their perspective "practitioners," equally beautiful) theories such as inflation failed to ignite such widespread interest (and tacit acceptance) by the general public? String theorist Amanda Peet (2007) of the University of Toronto offers her own explanation on her webpage of advice for students pondering a career in string theory:

> String theory is a subject full of "buzz": it encompasses many cool concepts, including quantum mechanics, relativity, extra dimensions of space, the arrow of time, the origin of the universe, and so forth. I think it's great that so many students are interested, but I'm a string theorist, so of course I'm biased.

Peet's unapologetic admission of partiality points us toward the heart of the matter: string theorists as a whole comprise a passionate and tireless cheerleading squad for their discipline, and as such have exploited the popular media in a manner that no other branch of science seems to have mastered.

Legendary science popularizer the late Carl Sagan lamented that "like some editors and television producers, some scientists believe the public is too ignorant or too stupid to understand science" (1997: 334). Add to this the unfortunate reality that many scientists are painfully incapable of explaining their research in an engaging and accessible manner. Nature abhors a vacuum, and those in the general public who have a genuine interest, but lack of formal education, in cutting-edge areas of scientific research have few choices. In the past two decades, two camps have rushed in to fill the void—string theorists and New Age/pseudoscience practitioners. Unfortunately, both groups often appear to speak in the same vernacular, and in many cases the average person has no means to differentiate between the two, as evidenced by the *New York Times* best seller's list, which has featured both *The Elegant Universe* by physicist Brian Greene and *Quantum Healing* by New Age guru Deepak Chopra.

By bringing their research directly to the general public through countless books, magazine articles, and television programs, always done in an enthusiastic (or, as some have claimed, an overly-positive and uncritical) and audience-mindful manner, Greene and fellow "stringers" have successfully painted their particular corner of theoretical physics as the "in-crowd," the veritable "A-listers" of the scientific world. Even high school and undergraduate science teachers seem to have succumbed to the superstring Sirens' song, resulting in an undergraduate-level textbook on the subject, and a conference organized by the Institute for Theoretical Physics (University of California, Santa Barbara) to train high school teachers (Woit 2002). To those in the scientific community who view the results of string theory with a more critical eye, this has become ever more alarming, as evidenced by recent popular-level criticism by Lee Smolin (2006) and Peter Woit (2006).

An enthusiastic consumer of popular-level articles and books on theoretical physics, science fiction/fantasy illustrator and artist Hannah Michael Gale Shapero pondered the popularity of such works on her blog and came to the following insightful conclusion:

> I just can't get enough of this stuff. In fact, it's so attractive to me that I might as well call it as I see it: it's *science porn*. (March 17, 2004).

In her review of physicist João Magueijo's *Faster Than the Speed of Light,* a popularized account of his research into the speculative theory of a varying speed of light (VSL), Shapero explained the aspects of this literature that she found paralleled pornography as it is commonly understood, such as how it is filled with:

> seductive words and stylized, unattainable people and breathless scenes of discovery-consummation. It's that the whole enterprise (sorry, Dr. Krauss) of science is made to sound sexy and exciting...Invariably, the author apologizes for not putting mathematics in his book. But if he did, it would cause the enthusiasm of the potential reader to droop. It is part of the ritual of pornography that the encounter fantasy is made artificially easy. The climax of discovery gets a lot more pages than the long years of struggle that preceded it. To someone excited by science and natural phenomena, the tales of galaxies, black holes, parti-

> cle explosions, string theory, and esoteric cosmology are intellectual porn. You read it for the pleasure and the excitement, you want to be with it, and you keep wanting more...But there is also an underlying element of futility. The science-titillated can gaze and read and look and fantasize, but they cannot have it, cannot be part of the scientific work or make discoveries themselves, because for the non-mathematical public, popular science writing will always be a carefully presented illusion...(*Ibid.*)

This excitement of discovery has always driven scientists, including theoretical physicists, in their quest to understand the natural world, and is often described using sexual language. For example, physicist George Uhlenbeck noted that "I felt a kind of ecstasy" when he first heard of the Kaluza-Klein theory of fifth-dimensional unification in 1926 (Smolin 2006:47). Carl Sagan, whom Shapero (March 17, 2004) calls "the previous generation's science-porn-celebrity," admitted to feeling "a tingle of exhilaration" (Sagan 1997:330) concerning scientific discovery. In fact, it was his life-long love affair with science which drove him to write for the general public, writing that "*Not* explaining science seems to be perverse. When you're in love, you want to tell the world" (*Ibid.*: 25). Fellow science popularizer Stephen Hawking, whose *A Brief History of Time* is undoubtedly the best-selling work of "science porn" of all time, openly offered that "there's nothing like the Eureka moment, of discovering something that no one knew before. I won't compare it to sex, but it lasts longer" (2003:117).

Magueijo makes an equally sexual reference in *Faster Than the Speed of Light* when describing Edwin Hubble's discovery of the expansion of the universe, done "without 'attaching' the naked eye to the end of his telescope, using instead photographic plates that could be exposed for very long periods. What came out of these unusual observations was truly pornographic" in its shocking detail (2003:74). Shapero cites this incidence of sexual language and the plethora of others in his book, and finally criticizes Magueijo's language (which one might deem to be more suitable for a sailor than a scientist) as being over the top (March 22, 2004). For example, he describes specialized research facilities as "wank-facilities...a substitute for sexual intercourse for the aging scientists" (253), and notes that quantum gravity theorist Lee Smolin had a healthy viewpoint concerning his area of research because unlike his colleagues he "did not believe that God was about to give him a blow job" (247). It appears that in the world of science porn, as in the film industry, "soft-core" sometimes attracts a wider audience as opposed to "hard-core."

String theorists have generally walked this line with success in their popular writings. Brian Greene, undoubtedly the most successful of the stringy "science-porn-celebrities," describes the "passionate drive to understand the origin of the universe" and "passionate scientific investigation...about the nature of the universe" (1999: 345; 2004: ix). In his bestsellers we read about the impossible-to-resist "allure of string theory," as its practitioners "grope for an analogous principle that would capture the theory's essence as completely [as general relativity]" (2004: 352, 376). Greene attempts to convince the reader that since the extra dimensions predicted by string theory would "so profoundly influence basic physical properties of the universe, we should now seek—with unbridled passion—an understanding of what these curled-up dimensions look like" (1999: 206).

City University of New York physicist Michio Kaku, who along with fellow string and M-theory researchers Leonard Susskind, Brian Greene, and Lisa Randall have been dubbed "rock star physicists" (Kuahji 2006), has also turned his palpable passion for string theory into successful popular-level writings which qualify as "science porn." Kaku willingly admits to being "obsessed" with physics, a "source of joy and frustration" which is "the thrill of my life" (2005: 48). The idea that superstrings could be the unifying theory of everything is "intoxicating," and M-theory, its eleven-dimension extension, leaves "many theoretical physicists (myself included) gasping at its power" (Kaku and Trainer 1987: 4; Kaku 1997: 32).

Just as women writers, producers, and directors have made successful inroads in the production of literary and cinematic pornography, female physicists, too, have (in admittedly limited numbers) breached the largely male bastion of theoretical physics, including superstrings and M-theory. Amanda Peet calls string theory "seductive," partly because "it incorporates a number of really beautiful ideas in mathematics." She further explains that she and her colleagues pursue string theory "with such fervor" partly because "it's just a really sexy theory" (2003). She further utilizes a rather masculine (or at least Freudian) sexual metaphor in noting that "part of the reason that it is so sexy is...our tool kit is big" (*Ibid.*).

Lisa Randall, the first woman tenured in physics at Princeton (more recently affiliated with M.I.T. and Harvard), received the 2007 Julius Edgar Lilienfeld Prize from the American Physical Society for her work in theoretical physics and popularizing that research. Her thick 2005 work of "science porn," *Warped Passages,* featured copious references to such varied musical icons as Metallica, Cypress Hill, Jefferson Starship, Bjork, Eminem, U2, and REM. Randall's work in brane-models of the universe straddles several sexy topics in theoretical physics, including superstrings/M-theory and cosmology. Like her male colleagues, Randall repeatedly refers to science as her love, a topic that is "irresistibly exciting" and "thrilling" (2005: vii). Comparing scientific models with those in the fashion industry, she finds that both "demonstrate imaginative creations, and come in a variety of shapes and forms. And the beautiful ones get all the attention" (70). Interestingly, Randall also lauds the virginal nature of new scientific models, which are "untouched, delightful terrain" (8).

If Randall is correct, and beautiful scientific models garner the most attention, then we should be mindful of two trite yet true sentiments: beauty is in the eye of the beholder, and it is only skin deep. The same might also be said of pornography (or any other creative activity). String theorists as a group frequently cite the mathematical beauty of their "beloved" as reasons for their devotion to it. For example, Nobel Laureate David Gross observed of string theory as early as 1988 that "as the theory develops, more and more people are convinced by its beauty. It is a very beautiful theory even though it's only rather primitively understood, so it's likely to be even more beautiful once we understand it at a deeper level in the future" (Davies and Brown 1988: 149). Just as pornography or other sexually explicit art has been criticized for its emphasis on (sometimes exaggerated) physical attributes, string theorists' continual and emphatic claims for the beauty of their models have opened it up for serious criticism. Quantum gravity theorist Lee Smolin (Perimeter Institute for Theoretical Physics) warns that not only is the "beauty" of a theory (like a human being) difficult to quantify, but that past lessons have shown us that "a theory can be fantastically beautiful, fruitful for the development of science, and yet at the same time completely wrong" (2006: 45). Mathematician Peter Woit (Columbia University), another of string theory's recent vociferous critics, paints string theory's beauty as similar to the smoke and mirrors of the film industry, namely a "beauty of mystery and magic," the type of beauty that "may disappear without a trace once one finds out the magician's trick behind the magic or the story behind the mystery" (2006: 197).

Although Smolin and Woit have become superstrings' most well-known critics, largely thanks to *The Trouble with Physics* and *Not Even Wrong*, their respective 2006 popular-level critiques of the theory and its practitioners, other voices have risen in criticism of string theory and its sociology over the past few decades. For example, as early as 1988, particle physicist and Nobel Laureate Sheldon Glashow called string theory a "contagious disease—I should say far more contagious than AIDS," a disease which he had unsuccessfully tried to prevent from infecting Harvard (Davies and Brown 1988: 191). In his own scientific autobiography of the same year, Glashow asked the (largely rhetorical) question, should string theorists "be paid by physics departments and permitted to pervert impressionable students?" (1988: 335). The palpable distain some physicists publicly voice toward superstrings (and their extension into M-theory) is also expressed using the language of sexual "perversions." For example, João Magueijo asked if the unknown source for the name "M-theory" might be something other than the usually cited magic, matrix, mystery, or mother-of-all-theories, because "M for masturbation seems so much more befitting to me" (2003: 240). This rather graphic analogy parallels a much earlier criticism of mathematics (to which string theory is often compared) attributed to Nobel Laureate Murray Gell-Mann, namely, that mathematics is to science as masturbation is to sex (Woit 2006: 189).

It is important to note that the use of sexual metaphors is certainly nothing new in science. Indeed, even today we still often speak of "Mother Nature," a rather ancient and widespread concept throughout many cultures. However, the reductionist post-Newtonian world of Western science has inherited not only the standard scientific method, but also an unfortunate viewpoint of the relationship between science and its object of study. In *Reflections on Gender and Science* (1985), Evelyn Fox Keller described the accepted methodology of science inherited from Francis Bacon in sexual terms:

> it is "natural" to guide, shape, even hound, conquer, and subdue her—only in that way is the true "nature of things" revealed...Science controls by following the dictates of nature, but these dictates include the requirement, even demand, for domination...Not simple violation, or rape, but forceful and aggressive seduction leads to conquest...(36-7)

Cohn (1987) recounts the widespread use of sexual imagery of domination in the development of nuclear physics and nuclear weapons. More recently, string theorist Brian Greene expounded a similar view in his 2004 bestseller *The Fabric of the Universe,* where he notes that "nothing comes easily. Nature does not give up her secrets lightly" (470). Princeton's Edward Witten, widely regarded as the Einstein of string theory, voiced a similar opinion when asked what had originally attracted him to string theory: "String theory is extremely attractive because gravity is forced on us" (Davies and Brown 1988:95). If we accept the widely-cited criticism of pornography as the victimization of women for the sake of (mainly) male entertainment, Shapero's analysis of science popularization as "science porn" takes on a more ominous tone, whose ramifications will not be explored further here.

Among the critics of string theory, perhaps the most gentlemanly is Lawrence Krauss of Case Western Reserve University. Also a purveyor of successful "science porn," Krauss's 2005 book *Hiding in the Mirror* focused on "our love affair with extra dimensions," including string theory (1). Even his description of this love affair is more along the lines of romance than raw sex (such as in the case of Magueijo). This romance, like those in real life, is not without problems, and he notes that "physicists have been fickle in their intermittent love affair with extra dimensions, turning hot and cold as their whims and desires evolved" (140). He calls extra dimensions a "romantic notion," that is "beautiful," and "seductive," but he warns that this Sirens' song, like others in the history of physics, may ultimately prove to be false. For example, he reminds the reader of the case of Grand Unified Theories (GUTs) where physicists found that "following the first flush of romantic love invariably comes the recognition that the object of one's affections is not quite perfect" (162). Krauss is at his most playful with the language of sex and science when describing the development of the so-called heterotic string theory, whose name,

> Does not derive from the word *erotic,* but rather from the root *heterosis,* although there is also no doubt the model is kinky, both metaphorically and literally. Indeed, it is so imaginative as to be considered sexy by many theorists, which says something either about the model or about the theorists. (182)

Suggestive remarks about the sex lives of physicists might seem out of place at best, a non sequitur at worst. After all, one of the stereotypes most deeply ingrained in the public psyche is that of the sexless science nerd. Carl Sagan (1997: 382-3) eloquently painted the public's cumulative mental image of his colleagues as follows:

> Nerds wear their belts under their rib cages. Their short-sleeve shirts are equipped with pocket protectors in which are displayed a formidable array of multicolored pens and pencils. A programmable calculator is carried in a special belt holster. They all wear thick glasses with broken nose-pieces that have been repaired with Band-Aids. They are bereft of social skills, and oblivious or indifferent to the lack.

If string theory truly is the "theory of everything," is this possibly a case of the revenge of the nerds? In his review of Smolin's *The Trouble with Physics,* critic Jim Holt referred to this "cult of empty mathematics," and noted Smolin's reflection that possibly "many leading theoretical physicists were once insecure, small, pimply boys who got their revenge besting the jocks (who got the girls) in the one place they could—math class" (Holt 2006: 86).

However, the best revenge for a stereotype is to shatter it by example, and the purveyors of "science porn" have successfully done just that. Shapero notes that in this genre "young male scientists such as Brian Greene and the movie-star-handsome João Magueijo are marketed in magazines and on TV as intellectual matinee idols" (March 17, 2004). Indeed, the entire back of the dust jacket of *Faster Than the Speed of Light* is devoted to a lovingly shot close-up of Magueijo's doe-eyed baby face, harkening this author back to pre-teen years guiltily spent pasting glossy *Tiger Beat* pin-up pages of Donny Osmond and Bobby Sherman on her bedroom wall. An article on *NOVA's* 2003 television mini-series based on Brian Greene's *The Elegant Universe* featured comments by Executive Producer Paula Apsell. In an almost gushing tone, she describes the photogenic Greene (who hosted the series) as a "sexy, smart scientist" and the project as "a marriage made in heaven" (Odenwald 2003). In his review of Greene's *The Fabric of the Universe,* Lee Smolin playfully noted that "I get mail from readers who complain that I am not as good-looking as Greene, even though I write better." Smolin is then quick to point out that his partner—"she-who-matters"—holds a different point of view (2004: 371).

The literary and visual media have emphasized the positive physical as well as intellectual attributes of other theoretical physicists over the past few years. In a 2005 article in *Cosmos Magazine,* writer Elizabeth Finkel tells her readers that Michio "stretches his mind to eleven dimensions, understands what Einstein failed to grasp, and he plans for the death of our Sun, five billion years from now. Michio Kaku is a superhero of the incomprehensible." In turn, Kaku describes the young and exquisitely beautiful Lisa Randall as "resembling the actress Jodie Foster a bit," and juxtaposes her to the "fiercely competitive, testosterone-driven, intensely male profession of theoretical physics" (2006: 217). In the 2004 *BBC/Learning Channel* special "Parallel Universes," theoretical physicists are highlighted taking part in athletic activities, in rather form fitting outfits: Michio Kaku ice skating, Lisa Randall indoor rock-climbing, and Neil Turok swimming.

Strange as it may seem, even the "physical prowess" of the most unlikely physics sex-symbol, "rock star physicist" Stephen Hawking, has been highlighted in public circles. String theorist Leonard Susskind calls Hawking (who has been confined to a wheelchair for over 25 years by Lou Gehrig's disease):

> the Evel Knievel of physics. Brave to the point of recklessness, Stephen is a well-known traffic menace in Cambridge, where his wheelchair is often seen careening around, way beyond safe speeds. His physics is in many ways like his wheelchair driving—bold, adventurous, audacious to the maximum. (2006: 326)

Popular-level articles frequently point to Hawking's affection for Marilyn Monroe, and often feature a picture of his disease-crippled body posing in front of a large poster of the sex-symbol, a beaming yet crooked smile lighting up his face. The fact that he fathered three children with his first wife, Jane, all after his diagnosis, is frequently mentioned, and in recent years, the soap opera nature of his personal life—from his divorce from Jane to marry Elaine Mason, one of his nurses, to his recent second divorce—has been the subject of international tabloids. In one of most widely-cited portions of her autobiography, *Music to Move the Stars*, ex-wife Jane describes how "of necessity" the couple's sex-life "was unadventurous and he was and always had been the passive partner because of muscle weakness," and further that as his physical condition deteriorated she increasingly found it "becoming very difficult—unnatural, even—to feel desire for someone with the body of a Holocaust victim and the undeniable needs of an infant" (Hawking 2000: 328).

As Kemp (1998: 551) explained "what a scientist (or artist, author, composer...) looks like should not matter to us...Yet we harbor an apparently irresistible urge to scrutinize the appearance of famous and infamous persons." Writers and producers in the "science porn" industry have nevertheless exploited this "urge" as part of their successful selling of science to the general public, ignoring all value judgments as to the morality of this urge in much the same way as producers of cinematic and literary pornography exploit the sexual urges of their audience. But Hunt (1996: 10-11) reminds us that pornography was born "as a literary and visual practice and as a category of understanding, at the same time as—aand concomitantly with—the long-term emergence of Western modernity" including the Scientific Revolution. It is therefore not surprising that after several centuries of tension and interplay with the larger culture, we find the genesis of a popular literature of science which shares characteristics with modern literary and visual "sexual" pornography. As popular as it may be, "science porn" runs the serous risk of painting an at best incomplete, at worst erroneous picture of science, just as pornography has been accused of fostering an implausible fantasy concerning sexuality. But lest this essay end on too somber a note, the author will offer one last sensationalized point: If string theory is indeed untestable (as critics charge) and therefore unscientific, then it may truly fit the legal definition of an obscene work, as established by the U.S. Supreme Court, namely, one that "A reasonable person would find…taken as a whole, lacks serious literary, artistic, political, and scientific value" (Wekesser 1997: 13). The reader is left to ponder the ramifications of this possibility.

References

Cohn, Carol. "Sex and Death in the Rational World of Defense Intellectuals." *Signs* 12.4 (1987): 687-718.

Davies, P.C.W. and Julian Brown, eds. *Superstrings: A Theory of Everything?* Cambridge: Cambridge UP, 1988.

Finkel, Elizabeth. "Fish Out of Water." *Cosmos Magazine.* August 2005. 26 December 2006. <http://www.cosmosmagazine.com/node/99>.

Glashow, Sheldon. *Interactions.* NY: Warner Books, 1988.

Greene, Brian. *The Elegant Universe.* New York: WW Norton, 1999.

---. *The Fabric of the Cosmos.* New York: Knopf, 2004.

Hawking, Jane. *Music to Move the Stars.* London: Pan Books, 2000.

Hawking, Stephen. (2003) "Sixty Years in a Nutshell." *The Future of Theoretical Physics.* eds. G.W. Gibbons, E.P.S. Shellard, and S.J. Rankin. Cambridge: Cambridge UP, 2003: 105-117.

Holt, Jim. "Unstrung." *The New Yorker.* 2 October 2006: 86-91.

Hunt, Lynn, ed. *The Invention of Pornography.* New York: Zone, 1996.

Kaku, Michio and Jennifer Trainer. *Beyond Einstein.* Toronto: Bantam, 1987.

Kaku, Michio. "Into the Eleventh Dimension." *New Scientist.* 18 January 1997: 32-6.

---. *Parallel Worlds.* New York: Anchor, 2006.

---. "Unifying the Universe." *New Scientist.* 16 April 2005: 48.

Keller, Evelyn Fox. *Reflections on Gender.* New Haven, CT: Yale UP, 1985.

Kemp, Martin. (1998) "Icons of Intellect." *Nature* 395 (1998):551.

Krauss, Lawrence M. *Hiding in the Mirror.* New York: Viking, 2005.

Kuahji. "Top Ten Rock Star Physicists." *The Hyperspace Forums.* 10 September 2006. 26 December 2006 <http://www.mkaku.org/forums/showthread.php?t=216>.

Magueijo, João. *Faster Than the Speed of Light.* Cambridge, MA: Perseus, 2003.

Odenwald, Dan. "*Nova* Strings Together A Theory That's Got it All." *Current.* 14 July 2003. 26 December 2006 <http://www.current.org/doc/doc03/3string.html>.

Peet, Amanda. " 'Straight Dope' Advice for Students Keen on String Theory as a Career." *University of Toronto Physics Dept.* 28 April 2007. 31 May 2007 <http://www.pep.to/physics/advice/straightdope/>.

---. "Viewpoints on String Theory: Amanda Peet." *Nova: The Elegant Universe.* July 2003. 2 January 2007 <http://www.pbs.org/wgbh/nova/elegant/view-peet.html>.

Randall, Lisa. *Warped Passages.* New York: Harper Collins, 2005.

Sagan, Carl. *The Demon-haunted World.* New York: Ballantine, 1997.

Shapero, Hannah Michael Gale. "His Mouth Goes Faster Than the Speed of Light." *Electron Blue*. 22 March 2004. 26 December 2006 <http://www.pyracantha.com/cgi-bin/blosxom.cgi/2004/03/22>.

Shapero, Hannah Michael Gale. "Science Porn." *Electron Blue*. 17 March 2004. 26 December 2006 <http://www.pyracantha.com/cgi-bin/blosxom.cgi/2004/03/17>.

Smolin, Lee. *The Trouble with Physics*. Boston: Houghton Mifflin, 2006.

---. "Unraveling Space and Time." *American Scientist*. July-August 2004: 371-3.

Susskind, Leonard. *The Cosmic Landscape*. New York: Little, Brown, 2006.

Wekesser, Carol, ed. *Pornography: Opposing Viewpoints*. San Diego: Greenhaven, 1997.

Woit, Peter. "Is String Theory Even Wrong?" *American Scientist*. March-April 2002: 110.

---. *Not Even Wrong*. New York: Basic, 2006.

New Institute at Stanford

Peter Woit

Stanford University will officially announce later today the founding of a new research institute, with major funding from the John Templeton Foundation. Many of the faculty and research staff of the new institute will come from the present Institute for Theoretical Physics, which will be shutting its doors.

Co-directors of the new institute will be Stanford faculty member Leonard Susskind, and Gerald Cleaver, who is currently head of the Early Universe Cosmology and Strings Group at Baylor University. Susskind, who is one of the co-discoverers of string theory, has in recent years been the most prominent promoter of the theory of the "multiverse," which he describes in a recent interview. Later this month he will be giving the Einstein lecture at Brown University on the topic of String Theory and Intelligent Design. He is widely considered to be the leading candidate for next year's Templeton Prize. Cleaver, a prominent string theorist who was a student of John Schwarz (the co-discoverer of superstring theory) at Caltech, has published more than 40 important research articles on string theory. Like Susskind, his recent interests have been in the area of string cosmology.

Next year the institute will open its doors with a year-long program on the topic of the multiverse, led by theoretical cosmologist George F. R. Ellis visiting from the University of Cape Town. Ellis, the 2004 Templeton Prize winner, explains that the traditional view of an opposition between faith and science has been made obsolete by the latest research in string theory and cosmology. Says Ellis, "In the end, belief in a multiverse will always be just that—a matter of belief, based in faith that logical arguments proposed give the correct answer in a situation where direct observational proof is unattainable and the supposed underlying physics is untestable."

The new institute will be named the Stanford Templeton Research Institute for Nature, God, and Science (STRINGS) and will collaborate with other related Bay Area organizations, including Stanford's own KIPAC (Kavli Institute for Particle Astrophysics and Cosmology) and Berkeley's CTNS (Center for Theology and the Natural Sciences). Steve Kahn, the director of KIPAC, welcomed the formation of the new institute, saying, "We're very pleased to have such a major institution on campus led by two such prominent physicists working on cosmology. In this era of declining NSF and DOE budgets, we need to branch out from traditional approaches to science. We expect to collaborate with the new institute to help us seek funding from sources such as the President's FBCI initiative." Besides the physicists, several faculty from other Stanford departments will be affiliated with the Templeton institute, including computer scientist Donald Knuth, author of the recent book *Things a Computer Scientist Rarely Talks About.*

According to Dr. John M. Templeton, Jr., president of the Templeton foundation, "the idea for the institute grew out of our involvement with a series of lectures at Stanford in the area of biology. At those lectures the biologists pointed out to us that it was the physicists on campus who were doing work most closely related to our foundation's interests, something we had already noticed through our Cosmology and Fine-tuning Research Program. As the latest cutting-edge research in physics has caused physicists to rethink what it means for a theory to explain experimental data, the wedge driven by Galileo between science and religion has begun to close. We're very proud to be able to support and encourage this trend."

Encouragement also comes from some other members of the Stanford physics department. Nobel Prize-winning theoretical physicist Robert McLaughlin was quoted as saying "theoretical particle physics is just getting old and losing its youthful good looks. Even Ed Witten has given up on it. This latest plan for the cosmology/multiverse/string theory crowd to join up with Templeton reminds me of a woman deciding to become a nun when she gets too old to attract men. But if it gets them out of the physics department, I'm in favor of it. Don't let the door hit you on the way out, guys."

Fictions

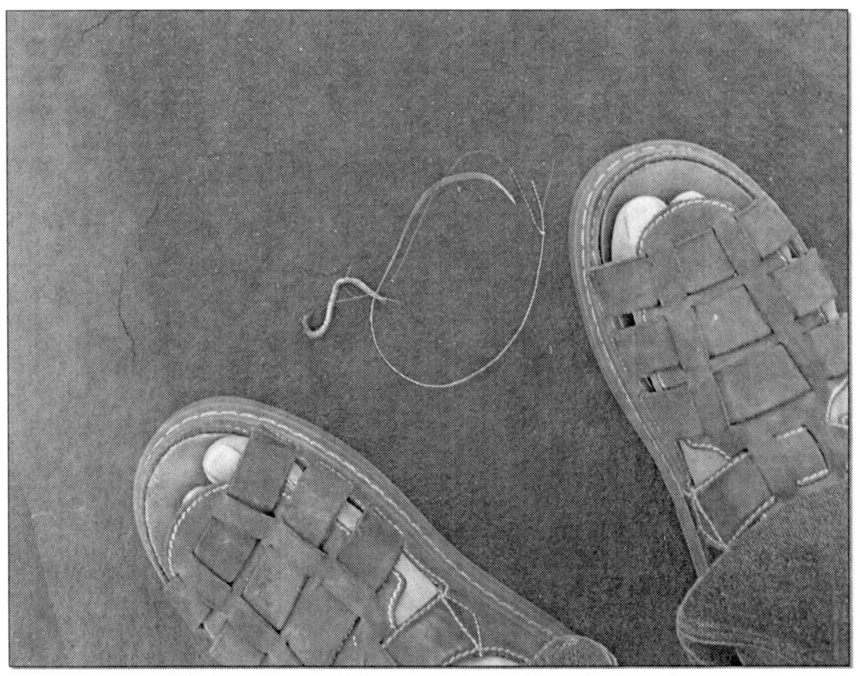

"String theory (vibrating)" by Alex Nodopaka

S-Bomb

Adam Roberts

What does the "S" stand for?

There's a black blotch in the sky where the starlight has been hoovered away. Any northern hemisphere night sky shows it. You'll have heard of this, of course. It can't be a planetary body, although it's round enough for that; but there are no gravitational effects detectable. One theory is that it is a concentration of dust occluding the starlight in a circular patch. There is concern, for the dust seems within the solar-system and therefore close to Earth, but it is below the line of the ecliptic and approaching no closer. There are of course plans to launch probes to examine the phenomenon. It's a question of finding the funding, of working out a launch window, that sort of thing.

**

—I'll tell you what. When they named the A-Bomb, they plugged into a cultural context in which A was the top school grade, and *A-OK* and *A1* had wholly positive associations. Even the word *Atom* connoted focus and potency, think of *the Mighty Atom*. And then, only a few years, the world hears of a *more* powerful bomb, the H-Bomb, and "H" meant nothing, except itself: Hydrogen. It connotes the gaseous, diffuseness, the whiffy. In their *heads* people knew this bomb was more deadly than the former, but in their *hearts* they couldn't truly credit it. So, I guess what I'm saying is, what, really, might people make of S? S-Bomb?

—*Sex-bomb.*

—Wasn't that a song?

—*If it was?*

—When I was a child, there was a pop group called S-Club. Or was it S-Group? But, see, S-Group, no. That sounds more like a secret arm of the military. I can't believe a kid's pop group would go for that sort of name.

—*And when* I *was a child, there was a pop group called the Incredible String Band. So what do you think of that?*

—The Incredible String Bomb?

—*Incredible, after all, is a pretty good word for it. From where I'm sitting I'd say that incredible describes it pretty well.*

—Except—these are no ordinary strings, Super, after all.

—*String bomb sounds like a Wallace and Grommit device. A back garden shed concoction.*

—See, that's my point. S-Bomb is a phrase that lacks the necessary.

—*Or further back? There's the echo of SS. No? The SS-Bomb? Some Nazi artifact. That sounds pretty mean.*

—Better. Better.

—*Also—I mean, you correct me, you're the expert—who calls them superstrings any more? Clumsy and rrropey metaphor.*

—I guess. S for Sub-materialities. S for Severe. Serious, ser, Seriousnesses. Sparks. Sparkles inside everything, and this bomb harnessing that.

—*Except, see if I understand right, not so much* inside *as—*

—Not inside things. No. Constitutive of things. Yes.

—*You do sound—nervy. Do you have something to tell me?*

**

The two of them were sitting in a coffee shop, the Costarbucks Republic, the Coffee Chain, whatever. There were two thicklipped porcelain mugs, large and round as soup bowls, on the table before them. Inside one a disc of blackness sat halfway down, with little pearls of reflected brightness trapped in its meniscus. The other mug was as yet untouched, and brimmed over with a solid froth of white that was dirtied with brown-black like pavement snow.

So much for their *coffees*.

The one man was old, a face like the older Auden, his nose fattened with age, two wide spaced inkdrop eyes. His hair was white and closetrimmed and expressive of the undulating contours of his big old skull. The other man was young, and you might call him handsome if you happen to find male beauty in that block-faced, pineapple-headed muscular type. But he was very nervous indeed; very fidgety, and anxious, and gabbly. Why was he talking about long vanished pop groups and suchlike chatter?

The place was partially occupied, readers and laptop-tappers distributed unevenly amongst the darkwood tables. Behind the counter two slender men, both with skin colored coffee-au-lait, waited for the next customer. It's a neutral place to meet, is the point of it.

**

What's the weather like? Aren't you interested? Look through these wall high plates of carefully washed and polished glass. What can you see?

It's a pretty windy day. The weathermen didn't foresee that. There have been clear-sky gales to the west. A weird turbulence, unspooling tourbillons to the north and the south that resonate into unseasonal storms, flooding, wreckage. Nobody can explain it. But it's only weather.

**

The two men sat in complete silence, the older one staring balefully at the younger, for two minutes. Two minutes is a very long time to sit in silence. Try it. Life is hurry and bustle. People come into the coffee shop and grab cardboard tubes of hot black and rush out. Those cars lurching forward, slowing back, lurching forward, slowing back, all day and every day, such that the tarmac is being continually obscured and revealed.

The moon appears no larger at the top of the skyscraper than it does on the ground. The sun moves through the sky. But it doesn't. It's the sky moves around the sun. That's the truth of it.

The older man sat upright, and his little felt-circle black eyes seem to expand. Those white fur eyebrows, up they go, towards the hairline.

**

—*Run me through again what I am to tell my bosses.*

—Well, sorry, is one thing.

—*We'll take sorry as read. We'll assume it.*

—Obviously we should have been in closer communication with—by we I don't mean *me*, specifically, individually. We're a team, obviously—but, see, I'll be frank, *scientists*, our first reaction is, wait and see. It'll be OK. We think we can sort the problem, present you people with problem and solution in one neat package. Or at least, wait until there's a proper quantity of data before we report anything.

—*You saying there's no solution?*

—No.

—*You're saying the bomb doesn't work?*

—

—*I take your startlement as a yeah.*

—Sorry—sorry—you think that's what I'm here to report?

—*As opposed to?*

—Oh, the bomb works.

—*You're sure?*

—We tested it.

—*You have already tested it?*

—Tested it. It works. Jesus.

—*The people I work for will be pleased to hear that at any rate.*

—What I mean is. Look.

**

You think superstrings are myriad little-little separate strings, one-dimensional extended objects that resonate and shake, that aggregate and disaggregate into subatomic particles, and thence into atoms and molecules and everything in this diverse and frangible world. You think so. Think again. Think *laces*. Think of it this way: one single string, ten-to-the-million meters long, weaving in and out of *our* four dimensions, like laces weaving in and out of cosmic fabric, tying it together. Superstrings is a misnomer. This singular thing, this superstring. The equations require ten dimensions, and we're personally familiar with four dimensions, and all that is true. But when you look at it clearly, there *is* only one dimension. Only the one singularity, the thread that ties all of reality together and also the thread out of which all reality is woven. The one string.

—One string.

—The nature of the technology is that, and the, the *thing* is, said the younger man.

—You're saying you broke it.

—I'm saying, said the younger man, and swallowed air.

The older man lifted his coffee mug, finally, and tucked his white moustache into the white cap of froth.

—S-Bomb, boom-boom, said the younger, and the explosion. Now we were surely not expecting the explosive outgassing, the violent rupture, the A-Bomb thing. But I *was* expecting—I don't know. Maybe sparks, the sparkles, something fizzy.

—None of that?

—Then, said the young man, it detonates. The point is—you're wondering if I'm going to get to the point. The point is, it blows, but not with any explosive detonation. These strings, these threads, these laces stretching, as it were, across ten dimensions, connecting it all together, the whole of reality. Cut them, and, plainly put, our dimensions start to *unweave*, or unspool, or unpick, you choose the *un-* word you like best. It's a baseline reality event. The earth turns away from it. That's not a metaphor. The earth turns; it spins around the sun; it leaves the event behind at the speed of kilometers a second.

—You tested it underground?

—On the contrary. We tested it in the sky. We lifted it up there by a toroid helium balloon. No, no, if you dug it *under* the ground ...

—Let's say, interrupted the older man, under Tehran. Under and a little east of Tehran.

—Sure. Then the world itself moves through space, and the effect is to blast out an empty conic up from underneath the city. A hollowness that shoots out, angled and up out of the city and goes into the sky at a tangent, and loses itself in space, the city thereby collapsing into a great mass of rubble. The air, meanwhile, rushing about to fill the vacancy. But it's gone in minutes, because *we all* are travelling at such prodigious speeds, because the world is in orbit about the sun.

—So you're saying that, in effect, the point of detonation of an S-Bomb will appear, from where we're standing, *appear* to hurtle away up into space, said the older man.

—Yeah. Or it might cut a tunnel right through the earth, depending on the world's orientation when it was detonated. Or it might just fly straight up. The earth orbits the sun at about 30 kilometers a second. The sun is moving too, with us in tow, and rushing in a different direction at about 20 kilometers a second. That's a fast sheer vector. It means that the blast *leaves* the world behind pretty rapidly, hurtles above the plane of the ecliptic and away.

—And now you're going to tell me, said the older man, speaking expansively, a voice expressive of confidence, that the vacuum of space *neutralizes* the effect. It just burns itself out up there.

—See, said the younger man, leaning forward, we *wondered* about that. One S-Bomb theory was that, without matter to, to unpick, then it would just put-put and out. But the way it's turned out—no. It's expanding explosively. Faster than any chemical explosion, expanding really very quickly. But not so quickly as we are moving away from it, in our solar and Galactic trajectories, so in that sense we're safe.

—So what's the odds that our planet will swing round on its orbit into this expanding explosion, this time next year?

A weird little trembly high-pitched laugh.

—Man, no. What, the sun, you see. *Is* moving relative to the galaxy. And the galaxy is...anyway, it's a complex spirograph tracery, our passage through spacetime. So we're leaving it pretty far behind us, a spoor of vacuum-vacuum, unstitching the poor fourfold house in which we live. Like the wake of a boat. Or, from its point of view, we're skimming away as it swells.

**

—Now you're sure, said the old man, as he got to his feet, that it's a *real* effect?

—It's real, said the younger man. You know what? I'll level with you. We calculated a forty-sixty possibility that something like this would happen. Something like this. That's why we detonated it high in the air, so that the world would spin us away, day by day, and leave the detonation footprint behind in the vacuum. We figured, it's vacuum! What can happen? But it turns out, more than you'd think.

—So?

—Light propagates across a vacuum. Various electromagnetic radiations propagate across a vacuum. But none of them can propagate across the null space.

—Rubble can?

—What?

—You said, blow it under Tehran, Tehran falls into the hole.

—Well, yes. Because the earth swings away from it, leaves it behind. But, actually, weird things happen to the equations when you shuffle core assumptions about, you know, the fundamental premises of things. Atoms may tumble into null-space, but they get…churned. Or to be more exact: the Earth moves away, into a new baseline, and away from the detonation footprint, and then matter can move into the tunnel dug out by the S-Bomb. But they don't seem to, you know, stick as well as they ought. They seem to slide about more than you might think. But, anyway, at the point of the continuing detonation, *evidently*, electromagnetic waves aren't able to cross the null.

—So, said the older man, who is no fool, the black blotch in the sky. And all the pother in the media.

—And that's going to get worse. Nothing we can do about it. More and more stars are going to get blanked out by the phenomenon, in the northern hemisphere at least, in the backwash of the earth's passage through galactic space. Or actually the sun's, you see what I mean.

—OK, said the general. Long as it stays *out there*.

This is what he was thinking: biggest act of vandalism in human history. He's thinking: but leastways it's not pissing direct into our own pool. And as he extricates himself from the table his military mind is running through possible strategic uses, from attacking orbital platforms to high-altitude bombers, to maybe developing smaller or shorter lived devices that could be used lower down. He can't help thinking that way. He's a soldier.

—I had better go report right now, he said. My bosses will want to know this right away. Then, as an afterthought, *I'll* tell you what the S stands for.

—What?

—Starsucker. Starblotter. Or something (for he's never been very deft with a punchline) about stars. And he was at the door, and looking through the glass into the unseasonably windy weather. Go back to the institute, he said. Go back, and we'll contact you in due course.

**

This string, this one line out of which everything is spun, is broke; and the moment (the infinitesimal fractional moment) when that could have been repaired has long gone. Momentum works in strange ways in ten dimensions. Unspooling, unstitching, unpicking the tapestry of *matter* takes longer than unpicking the tapestry of *vacuum*. They slip free of their weave. The two whipsnapping ends of the superstring are acquiring more and more hyper-momentum. What does the S stand for again? Severed. Say-your-prayers. Stop.

**

Time continues applying its pressure and forcing the other three dimensions along its relentless and irrevocable line. For six months the coffee shop does its regular business, and customers come sluggish and drink and go off joyously agitated. There is a relatively high turnover of serving staff, for the pay is poor and the work onerous, but the two men who served during the conversation reported above are still in post six months on. Six months on is when the whole story breaks to the media: this cornpone country, its tiny research budget, its speculative endeavor, its helium-balloon-detonated-device unsanctioned by any international organization or superpower government. This devastation wreaked on the night sky (for the northern hemisphere night-sky is now a third blotted out by this spreading squid-ink), this hideous destructive power. Worse than atomics. The most massy of mass destructive possibilities. S for shock. Oh, the outrage.

The s-for-shit hits the f-for-fan. The government collapses. The country wilts under the censure of the international community. There's all that.

The whole story comes out. All the members of the eight-strong research team had been holding their peace under the most alarming threats from the security services; as had the dozen or so high-clearance security officials "handling" the case. But one succumbs, and defects, and reveals all; and then, one by one, so do the others. Some scurry up their local equivalents of Harrowdown Hill; some try and tease wealth from the media to tell their unvarnished tales.

For a brief period the coffee shop becomes a place of celebrity pilgrimage: it was in this very establishment, at this very table, that the scientific team first confessed their crime (and this is the term everybody is using now) to a security official. This is where the governmental cover-up began. Journalists, and rubberneckers, and oddballs, swarmed to the shop. The two men who had been on duty that day sold their stories; but their stories didn't amount to very much, and didn't earn them very much money.

But it is the nature of events that they entail consequences over a much longer timescale than people realize. The scientific community remains divided as to whether the unusually severe atmospheric storms are caused by the continuing action of the null-corridor, or whether the null-corridor has long since dissipated, and these storms are merely the long tail of the jolt which the chaotic weather system received from its initial carving.

**

And six months after *that,* the shop is empty. The small country that had produced this enormous device has been repudiated by many of the world's nations; there were economic sanctions in place, public shaming. It has offered up dozens of its official personnel, including all the remaining scientists on the team, to public trials and imprisonment.

—*Why were you so secretive? Why didn't you share the theoretical underpinnings of the technology you were developing?*

—We were a small group, working well within the budget for our team. The technology isn't expensive. The most expensive part of our equipment, in fact, was the balloon to lift it up for its trial detonation.

—*But* why *the S-for-Secrecy?*

—We figured we were like the Manhattan Project.

But, no! no! That doesn't wash. That doesn't wash.

—*The Manhattan Project was a wartime project. The secrecy was governmentally sanctioned, and a necessary component of the prosecution of the war. You were working during peacetime. You brought this horror on the world for* no *reason.*

—Not Manhattan Project in the sense of wartime, but Manhattan Project in the sense of knowing that we had a *potentially* catastrophically destructive technology on our hands. The last thing we wanted was for this to leak out. Our secrecy was motivated by a desire to protect humanity from the—

But it's no good. To prison they all go, for the term of their naturals, and the new government, and then the one that comes into power after that falls, makes repeated obeisance to the international community. And although some of its allies stand by it, the sanctions of others do bite. Its economy turns down. People lose their jobs. Poverty increases. It's all bad news.

Another government tumbles, tripped over by this immoveable object, this S-Bomb. Life gets harder still, and fewer and fewer people are in the position to afford frivolities like expensive coffee-shop steam-filtered coffee. The journalists are no longer interested. The ordinary disaster-tourists and rubberneckers don't call by any more. Only the weirdoes keep coming; and here's a truth about weirdoes: they're generally too parsimonious actually to buy the damn coffee. More often than not they come in, sit at The Table and run peculiar home-made Heath-Robinson handheld devices over it, up and down its legs, as if looking for something. Aluminum foil and cardboard and glued-on circuit boards and things like that, wielded as if the table could, if plumbed correctly, *reveal something* about the way an S-Bomb is constructed, or about the fundamental nature of reality, or things along that axis of thought.

**

The nations of the world, the ones that excoriate as much as those that stand-by, of course institute their own programs to uncover the technology at the heart of the S-Bomb. And it's not difficult, once you grasp a few general premises. Within the year there are a dozen functioning S-Bombs, none of which are publicly acknowledged. A year after that there are hundreds. There are different modalities and strengths of the device.

Does this sound like a stable situation to you? And yet another year slowly unspirals itself, and another, and another, without the world coming to an end.

The coffee-shop, to stay financially afloat, has rethought its business-plan to concentrate on cheap food, alcohol, and all-night opening. The expensive darkwood fittings and chunky chairs are starting to show wear and tear; and the clientele now mostly consists of people in cheap clothing who buy the cheapest soup on the menu, grab three breadrolls from the breadroll basket (despite the sign that says "*one* bread-roll per soup *please*"), cache two in their coat pockets, and then sit for hours and hours at their laptops trying to scratch together e-work. Thin chance of that, these days, friend. Hard times at the mill. They complain that the heating is turned down too low. The new manager stands firm. From next week, he decides, he's moving the breadroll basket behind the till. Customers will be issued with one roll when they have paid for the soup, no discussion, no argument.

But here's an old friend—looking no older. Close cropped white hair, whorled and scored skin. And with him, looking *much* reduced, the younger guy: thinner, raccoon-eyed, with a timid body language and a tendency to hang his head forward. And a third person: armed, e-tooled up with a head-sieve and fancy shades. The finest private bodyguard money can buy. He gives them privacy; checks out the space; waits by the door.

The two old friends can't sit at *the* table, since it was long since sold on i-Bay, but they buy some coffee and sit at *a* table, and that suits them just fine.

And for a while they simply sit there.

Eventually the younger one, his eyes on the tabletop and his manner subdued says: you taking me back after?

—Consider it your parole.

The younger man digests this fact.

—Not going back?

—No.

—I could tell you my opinion on the Antarctic business, he offers. This whirl-tempest thing. I have been thinking about it.

—We got people on staff who have been offering expert opinions on that.

This seems to pique the young man. I tried to keep up, he says, much as I could, as was possible within the confines of. But my internet access was severely, I mean *severely*, restricted.

—Really. Prison, says the older man. Who'd think it?

—What I'm saying (eyes still on the tabletop) is, I recognize that there will be people who have kept up with all the science better than I've been able.

—You're not out, says the older man, so that we can tap into your scientific expertise. That's not why you're out.

The obvious next question would be: then why am I out? But the young man has got out of the habit of interrogating others. So he just sits there. He keeps looking up at the bodyguard, flicking his eyes at the man's impassive face, stealing glances at the chunk stock of his Glock.

—Here's one thing, says the older man, you'll maybe have seen. Or heard about. The Chinese were trying to splice out a whole section of string. Best as I understand it, it would involve a double cut, liberating a continuous section, with the very rapid gluing-together of the remaining sections before they shot off to space forever at twenty-seven klicks a second. But the liberated section is carried along with us, apparently.

This gets the younger man agitated, although in a semi-contained, rather strangulated manner. See, this talking of splicing is a lie. You can't splice the string. The best you could do would be a temporary field-hold, and the equations include chaotic elements when you try and work out how long the hold is going to last. Not that you could do anything after the fact. If it breaks it'll be millions of kilometers away by the time it does.

He dries up, glances at the dour face of the older man, and then back at the table.

—Anyway, he says, in a gloomy voice. If you cut the string twice you'll *get* a continuous section. You just won't be able to say how long. It loops through ten dimensions, don't forget. It passes through six dimensions we can't even see. It might be a few meters long, or thousands of light years.

—A continuous whole section of that length, says the older man, drily, wouldn't be much use to us.

—But because it loops through so many dimensions…

—You think I don't know all this?

The young man looks up again, alarmed. Then, eyes down, he picks up his coffee and slurps it.

—This is the weave underlying everything, says the older man. We've all become pretty expert in this subject. This is the *ground*, the paper upon which the ink of reality is laid down, against which it is readable. Not only our world, but the whole cosmos, all matter and all vacuum, it all rolls itself along this endless medium; and without this medium it wouldn't exist as cosmos, matter, and vacuum. Everything material is relative, but this—this is absolute.

—I give the world, says the young man, one year. I'm amazed it's lasted as long as it has. This south polar sea incident—that shows you something. That shows you that S-Bombs *will* continue to be detonated. They'll be set off, by governments or terrorists, rogue states and idiots, and each one will knock another hole in the reality upon which we depend. Soon there will be hundreds of loose ends in the superstring. It will unspool more and more rapidly. It'll fray more and more.

—As I say, says the older man. We got brighter and better informed experts working for us now. Brighter than you, and better informed than you.

The younger man takes this in his stride, as how could he not? Seven years of prison are enough to break most people. He even nods.

—Let's say, the older man continues, that the Chinese have achieved this thing. We're not sure if by luck or judgment; but say they cut loose a segment of the unitary superstring. Say they unlaced it from ten dimensions into one dimension. One of ours.

—You mean, two?

—Just length. As breadth- and depthless as it is timeless. Or, let me be more precise. When it's looped about itself, or knotted, then effects of breadth and depth and time and other stuff are measurable. It's the *proximity* of one length of string to another length, and the precise pattern or orientation, of that proximity. One portion lying close alongside another, and you've breadth. Lying alongside another at a different orientation and you've time, and so on.

—They can manipulate it?

—So it seems.

—How? How can they? How?

—Their glue is better than our glue, I guess. They haven't created a discernable breach, for instance, so we think that they've found a way of holding the two severed ends of string in something approximating stability. They're in orbit, by the way, so maybe that helps. But our sources suggest they've got a separated out, whole, workable two-meter piece of string.

—That's very, says the young man, and he means to add, impressive, but the words dry in his throat.

—You know what they've found?

—What?

—The operation of this thing?

—I don't know.

—You couldn't guess. And we're not sure, because this is not first-hand. But by all accounts, the American security services, and ours, because ours depend upon theirs. This is what we've discovered: by manipulating it they manipulate grace.

—Grace.

—Grace, says the old man, and with this third iteration of the word he sits back in his chair and smiles. The curlicue grooves of his face buckle and chew, and his smile grows broader. It is alarming.

—I didn't realize you were religious, mumbles the young man.

—You didn't realize very much, returns the older, placidly, when you started on this project.

The young man looks up from the table, and there's a small flash in his eyes. I didn't, he says, realize I'd end up in prison, for instance.

—So why do you think we've sprung you?

The young man drops his eyes again, and shrinks back into himself, but he replies, in a low voice: I should never have gone to prison. My team were scapegoats. We worked under ministerial license, and carte blanche, on a weapon's program. If it weren't for the cap (which is what the half-sky filling northern hemisphere blackness is now usually called) and the baying-for-blood media, and the ignorance of the public, then...

He stops.

—And anyway what, he asks, do you mean, *grace*? Grace? What's that?

—You know, says the older man, turning his right hand over and back and over as if signaling 'so-so' very slowly, Grace. Beautiful sunsets. That lovely tickle inside your chest on Christmas morning. The tremendous mystery.

—What are you talking about?

The old man sits forward, and his deep wrinkles settle on his face. Oh, he's serious *now*.

—The medium of matter, the medium which enables the plenitude of the material. You know what the S- turned out to stand for? Spirit. That's what we've been dabbling with, cutting and splicing. And the Chinese, by all accounts, have made a machine that includes a one-dimensional stretch of Spirit. And who knows what they can do by manipulating it? Do you think they can kill or heal? Bless or damn? Some of the reports are pretty hard to credit, actually. But it won't stay under wraps for ever. These things never do.

There is a little more color in the young man's face as he looks up.

—You always knew, he says, that we had been specifically tasked with developing an S-Bomb. The orders were sanctioned from the highest levels of government. We were doing what we were told to, up to and including organizing tests. And then, when public opinion went sour, we were the ones hung out to dry. How many of my team are there even *left*?

—Atonement, says the older man. That's why you're out, now. That's what we're preparing for. Sacrifice, atonement. Transgression and forgiveness. We're working on the best information we have. But these are going to be the materials of the new dispensation.

—Whose transgression? says the younger man, sharply. Whose forgiveness?

—Another thing not in the news. A certain...organization...claims to have sunk a working S-Bomb into the Atlantic east of the USA. If they detonate it at the right moment it'll rip at twenty-five kilometers a second right through the world. It'll set off catastrophic earthquake, oceanic storms, it'll froth up atmospheric turbulence such as the world has never seen, before we leave it behind in space. We're in negotiations with them about the sums of money they want not to do this.

The young man is looking at the table again.

—You understand what I mean?

Nothing.

—You think you have suffered? says the older man. You think your sufferings have even *begun*? This discovery, and these weapons, belong to a reality whose laws we understand in only the crudest way. But if its currency is atonement, then who is better placed to offer himself up to that than you? There have never been such dangers of death facing the world. Do you understand the ferocity of what you've done?

—We didn't mean...

—Ask yourself again: why have we brought you out of prison? Why would we? How can you help? In what way can you atone?

The young man stares a long time at the tabletop.

The older man leans forward, and speaks in a rapid, low tone, as if pouring the words directly into the younger man's ear. Listen, he says. Listen to me. It's always been this way with bombs, on the one hand the rocket that hammers cities to powder, on the other the rocket that elevates human beings to the moon. It's always been this way. Your little S-device has polluted a third of the night sky with opacity. Three more have been detonated now, spreading their ink. How could it *not* be the case that, understood properly, this same device will heal?

The young man, eyes down, keeps staring.

**

As they talk, the proprietor sits on a stool behind the reinforced cash register, reading the paper. This is the lead story: experts say S-Bomb death spreading through the universe. This is the gist of the story, in which "cosmological expert Jerry Lowell" is quoted:

> *If the universe were infinitely big and filled with an infinite number of stars, then the night sky would be white, because no matter which direction we looked out our line-of-sight would end, eventually, with a star. There would be interstellar dust, of course, which you might think would occlude the lines of light, but in an infinite universe these would heat up and incandesce. But we don't see a white sky when we look up at night. So perhaps the cosmos is finite.*
>
> *But what if the S-Bomb technology is, like mathematics and nuclear power, something that every civilization discovers in due course? What if there have been millions, or billions, of alien civilizations out there that have discovered the S-Bomb, and detonated them, and left behind billions of slowly expanding spherical blots of impenetrable blackness. What if the dark between the stars that we see when we look up is that...these inevitably unspooling spots of death, growing eventually to devour everything? What if that's the truth of it?*

Another customer, and the proprietor folds the paper away and gets off his stool. A white porcelain mug, and the nozzle squirts black coffee promiscuously. It covers the white circle at the base of the mug almost instantaneously.

Eula Makes Up Her Mind

Daniel Conover

To: sendoggett@congserv.us.gov
From: jhughes@intnatscico.inasa.un.gov
Via: encryp.net331/priority subultra.1521003/Wlserver_nfs.7143393/7.14.39

Dear Senator:

Thank you for the heads-up on this week's hearing, or ambush, as the case may be. We appreciate your support, and we need it now more than ever. Please know that all of us up here are rooting for you.

As for what you should tell the other members of the committee, if it were me, I'd tell them all to shove it in sideways and break it off, but off course that's why I'm up here and you're down there in the political realm. Instead, I suggest you remind your colleagues that Delphi is a zero-gravity biological computer that is programmed for self-awareness. You don't just boot it up and start typing. We're as frustrated as anyone down there, but our circuit mapping suggests that Delphi's processing patterns are getting more elegant and less chaotic by the week.

Senator Beasley's complaint that continued black-budget funding for Delphi should be axed to appease the Chinese delegate is to be expected, but it's nevertheless pathetic. While those of us on-station are not directly affected by the climate and population problems, we're not completely ignorant, either, and we all hope to come home someday. If the U.N. is still committed to the colonization option, then the delegates must be reminded that the key to constructing interstellar quantum drives lies in the ability to do math in 10 dimensions, as dictated by string theory. Without a computer that can conceive those aspects of space-time, much less solve the engineering problems, we're going to be trapped within the tyranny of fuel-to-weight ratios for the foreseeable future. This is a surprisingly simple piece of logic, but then again, we're dealing with politicians who can't even do math in four dimensions. No offense.

Please make it clear to the committee that while the Delphi Project may be behind "schedule," it is not a project in disarray. We all remain committed to its goals and we are "working the problem." Morale is high, and we're making progress.

As for the senator's assessment that our interface cannot be achieved, please inform the committee that Pez will be joining us here via the next shuttle. His report on human interface with artificial intelligence (nsa_memo_Rf4/02/39_arch-

No.238_32_pb51.2) from Huntsville should hold everyone off long enough to give us a chance to test his theory. We have full confidence that Pez's approach will unlock new doors to the mystery of Delphi's consciousness.

Give 'em Hell on Tuesday, Senator...Jim

# # #

Ed "Pez" Pezzoli couldn't help but be impressed by how much progress had been made since the first time he'd come to Charleston, South Carolina, just three months before. On the earlier trip, he'd sat for an hour waiting for the ferry to carry him across the shallow inlet in what used to be called the Charleston Neck—the most narrow point in the peninsula. The bridge was under construction then, but the work was finished now and he glided over the developing marsh on a smooth ribbon of concrete and asphalt. In another ten years the debris-studded mud flats might look like they had been marsh forever.

He had found Eula Manigault in a sorry state on the first trip. She had answered the door in a bathrobe, staring at him with dishwater eyes. Reaching through the depression to her had been a task, but once they connected the rest was stunningly simple. Eula was mourning the death of her only child, but she didn't really want to lay down and die. She just hadn't found a better alternative.

Once his car whined down into Old Charleston, it was just a few minutes to Eula's room at the Omni. Nothing was very far from anything on the diked island, and the Omni lay in the center of it all, a five-star hotel in the midst of a museum city. Pez switched to batteries as soon as he came down off the interstate, and the valet was at his door with an electric plug as soon as the hybrid whirred to a stop at the Omni's grand entrance.

"Mister Pezzoli?" the valet asked.

"Yes."

"Miss Manigault will be right down. You wanted a quick-charge, right?"

Pez nodded, handed the valet his swipe card and peered through the smoked glass into the lobby. It surprised him how much he'd missed her in just a week. Eula was like that: Give her five minutes, she'd make you think about her for the rest of your life.

Over the three months they had spent training for the mission in Huntsville, Pez had watched Eula transform from a billowy zombie into a vigorous Earth mother. Her once-floppy arms now shined with muscle, and her body—still thick and motherly—had gone from flabby to feminine. Pez congratulated himself for following his gut instinct. The basket case with the off-the-charts synesthetic abilities had proven to be one of the most resilient human beings he had ever met. Maybe tragedy makes us stronger, he thought. Maybe someday I'll learn that.

He smiled and waved as she came through the door. Eula had put her hair back in a bright African scarf and Pez thought she looked surprisingly pretty. Eula dispensed with the formalities. Eula always dispensed with the formalities.

"Pez, you little cutie, I have had the best week!" she said as she enfolded him in a smothering hug. Her head only came up to his throat, but she almost squeezed the breath out of him. "I want you to meet my family!"

There was Aunt Shirley, who had raised her, some more aunts, some uncles, some cousins, and ebony platoons of nieces and nephews. They spilled out of the air-conditioned lobby into the wilting July heat, surrounding him, shaking his hand. Aunt Shirley hugged him, squeezing him hard the way Eula had, and she strained up to kiss his cheek.

"Pez, you all been so good to my baby. She's just so excited to be going into space, she can't hardly stand it."

"Your niece is very important to the work we're doing, Miss Shirley."

"Well, you just take good care of her, Pez. I done trusted you with her, because you're a nice man." Some steel entered her voice. "But you remember, now—if anything happens to my baby, I'm gonna come looking for you."

Pez sensed that would be a bad thing.

"Don't go scaring Pez, Shirley. I'm grown."

The two women hugged, but there were no tears. Just incandescent smiles. After everyone took their photos and the valet unplugged the hybrid, Pez loaded Eula into the car, everyone kissed one more time, and off the two space travelers whirred, with Eula's skinny, toothy nephews running alongside like chase planes.

"Have a nice leave?" Pez asked as she rolled down the window and lit a filtered cigarette.

"Delicious," Eula said, exhaling the smoke into the heavy Charleston air. "I want to thank you for putting everyone up at the Omni. Aunt Shirley had never set foot in a five-star hotel before."

"Don't thank me. Thank the black budget of the International Science Council."

"Well, it's about time the black budget started helping black folks, if you ask me. Did you have a good trip?"

"Oh yeah, " Pez said. "Said good-bye to everybody."

"With a history like yours, that could take a while."

"Took me about five minutes," he said. "I spent the rest of the time watching old movies."

Eula laughed and punched his shoulder. Pez accelerated as they climbed the ramp over the dike onto the interstate, switched over to the fuel cells, punched in the trip data, and relaxed into the plush driver's seat for the five-hour trip to the Cape.

They were almost to Savannah when Eula broke the drowsy silence.

"Aunt Shirley went to that root doctor in Givhans again, right before I left. He told her this whole business is being arranged by Anthony. Don't that beat all?"

Pez smiled and nodded.

"He was an amazing little child, Pez. But a dead boy arranging for his mama to go into space to talk to a computer? I think that root doctor has gone around the bend."

"It could happen," Pez said. Eula scowled and wagged her finger at him.

"Don't you patronize me, you skinny little computer geek, or so help me I will snap your neck like a pencil, that's what I'll do."

"I'm not patronizing you. I just don't make a habit of disagreeing with beautiful women who hold my professional future in the palms of their hands."

"Beautiful?" She smiled again. "Well, in that case you're a very handsome, wise man, and I take it all back. But now put on some tunes, honey, because you are just painful boring to ride in a car with."

They listened to Miles Davis the rest of the way.

# # #

To: sendoggett@congserv.us.gov
From: jhughes@intnatscico.inasa.un.gov
Via: encryp.net331/priority subultra.15210037Wlserver_nfs.7143393/7.24.39

I have to ask this, and please don't take offense, but do you think Senator Beasley's projects have to answer to this kind of scrutiny? Frankly, I doubt it. These are scientific matters, and the details the committee is asking for are going to be completely meaningless to laymen. So I will try to respond in a way that you can communicate to your meathead colleagues. If only there were a few more men with your vision in the governments of Earth, we wouldn't be in this fix we're in now!

Question No. 1: Professor Pezzoli is entirely qualified to do this kind of work. He has an engineering degree from MIT, and a degree in psychiatry from Johns Hopkins and he has written numerous scholarly articles on the subject of direct-to-cortex interfaces, not to mention last year's classified study for the Pentagon. The suggestion that Dr. Pezzoli is some kind of quack is not only insulting, it's absurd.

Question No. 2: Synesthesia was not discredited during the first decade of the century. Rather, synesthesia was discovered late in the 20th century and the phenomenon was studied sporadically for the next decade. Since a practical application of the knowledge gathered about synesthesia could not be divined at the time, funding dried up and serious research was discontinued.

I have attached several links to scholarly articles, each of which concludes that synesthesia is a physiological condition that begs for further study. Since it is not widely understood, let me clarify what synesthesia is so that you can answer any questions that may arise on the committee. Synesthesia is a relatively rare condition (one in every 2,000 Americans) in which neurological pathways that are universal at birth remain open into adulthood. These pathways are associated with accelerated infantile learning, but they close off about the time that language development becomes pronounced. Synesthetes in childhood and adulthood experience sensory connotations to words, names, concepts, numbers, etc. For instance, you may tell Senator Beasley that the letter "B" is "seen" as a brown oval with a green outline by 53 percent of all synesthetes (Pezzoli, 2035). You may also tell Senator Beasley that the word "idiot" is seen as a yellow-green rectangle on a brown background by 36 percent of the test subjects. Early research suggested that synesthetes may have additional capacity for learning, which should infuriate your esteemed colleague.

Question No. 3: Pezzoli's interface model has worked in laboratory tests. As you know from the Pentagon study (I cannot attach it here for security reasons), direct-to-cortex control is not only possible, it has been achieved. The same side effects noted in that study apply, but Pezzoli has controlled for those effects. Both of the women selected for the Delphi Project are well aware of the safety issues and have logged dozens of hours linked to computers that model Delphi-like behaviors. All ethical parameters in U.N. Protocol 251 have been documented (see attached).

Question No. 4: Pezzoli contacted his subjects by posting a request on a synesthete list-serve in the winter. Applicants submitted detailed profiles, and the ten selected for further study were chosen from a pool of the most highly educated respondents (our theory is that a successful human interface will have a trained and highly structured mind, in order to interpret Delphi in a meaningful way). Subject No. 1, Brittany Reynolds, is a medical doctor, trained at Duke University. She is 28 years old, comes from a stable environment and is considered our best interface candidate. Subject No. 2, Eula Manigault, did her undergraduate work at Howard and earned her PhD from Columbia. While it is true that her areas of study are English Literature and philosophy, we (Dr. Pezzoli) believe that her synesthetic gifts more than compensate for this weakness. Both women are single and childless.

Conclusion: While there is no doubt that attempting an unplanned human interface with Delphi is a somewhat unusual step, it is not "the desperate flounderings of a drowning project." Neither is it true that our plan is to "plug a couple of modern voodoo visionaries into an advanced computer" and then "try to read the tea leaves afterward." You might try telling Senator Beasley that the reason it's called "advanced science" is that if it weren't advanced, people like himself might be able to understand it.

On a personal note, Ron, your last message bothered me. Don't give up hope! We've been up here on Delphi for two years and we're closer to an answer than ever before. To cancel this project now would be like recalling Christopher Columbus just before landfall. Stick with us, Ron. Go down in history as the brave statesman who followed his great vision and fought the good fight and enabled mankind to finally become a people of the stars.

Yours in faith and perseverance,
Jim

# # #

Brittany's first trip inside Delphi lasted about thirty seconds. Her uncontrolled screaming lasted two full minutes.

"So much for option No. 1," Pez muttered to Jim Hughes in the narrow corridor outside the sick bay. Hughes was in no mood for the infamous Pez stoicism.

"Don't give me that. I sank a full third of this year's budget into your proposal. I put off two competing proposals entirely. If you're telling me that all your trials were flawed and that hooking a human up to Delphi is going to drive the subjects crazy, then you've got some explaining to do to the ISC. And if you're even thinking about pulling the plug on this, you're going to be a very unpopular man—which is not what you want to be on a space station."

"I'm used to being unpopular. Ask my ex-wives. And I never said anything about pulling the plug. Besides, we don't know that she's crazy. Maybe she's just…upset."

"Pez, she lost control of her bodily functions within 10 seconds. They're in there draining the sensory-deprivation tank as we speak."

Doctor Gherald stepped out of the medical section and closed the door behind him. He did not look pleased.

"Okay, which one of you is going to explain this to me?"

"It's my computer, but it's your interface, Pez," Hughes said. "Tell him."

"Brittany is a synesthete. We linked her cerebral cortex to Delphi via an alpha-wave translator—the same kind of translator we used in the trials back in Huntsville. And she just freaked out."

"Freaked out?"

"What, you want me to get more technical?" Pez protested. "She freaked out, man. One minute she's in a calm Alpha state, the next she's screaming and splashing around in the tank. We took her down immediately, but she just kept on screaming."

Gherald glared at him.

"You're lucky she's not dead. Do you have any idea how irresponsible this whole business is?"

"Oh for chrissake, Gherald, let's not make this worse than it already is," Hughes said.

"Did you test this system out on Delphi?" Gherald asked.

"This was the test," Pez said. "That's what we're trying to explain to you."

"Well, gentlemen, let me explain something to you. I'm the medical officer on Highland Station. I can't shut down your program, but I can by-God put Brittany on indefinite bed rest. So if you want your subject back, you're going to have to give me a better explanation than 'She freaked out, man.' Got that?"

Hughes intervened, laying a friendly hand on Gherald's shoulder, but Pez spotted the vein bulging in his forehead.

"Look, Doc, we'll get this all sorted out. Most likely the translator was improperly calibrated. Okay?"

Gherald stared at them skeptically.

"Listen, what I really want to know is when you're going to let me in to see her," Pez said. "She may be my subject, but she's also my friend. Is she conscious?"

"Yeah, she's conscious. She's in there with Eula now, and she's mildly sedated. You can see her, but if you upset her in any way, I am going to be...displeased."

The three men entered the sick bay, where Brittany lay with her head in Eula's lap, letting her friend stroke her short blonde hair. Eula smiled at them.

"Our baby is gonna be okay," she said. "Ain't that right, baby?"

Brittany reached up to stroke Eula's arm, but it was a listless movement, more distracted than intentional. Pez bent over and held Brittany's hand.

"How are you?" he asked.

Brittany scanned his face as if she were trying to place it.

"Why did you leave me in there so long?" she asked.

# # #

"You don't understand the problem at all," Eula told Pez as they took their morning exercise walking the outer rim corridor of the wheel-shaped station. "It doesn't have anything to do with the interface. It has everything to do with Delphi."

"That's not what I'm saying. You're getting hung up on words."

"But that's it, honey. Words. If I plunged you into the mind of God, words would make a lousy lifeboat, wouldn't they?"

"Delphi isn't God. Delphi is a computer."

"Okay then, Delphi isn't God. But don't go telling me it's just a computer."

Pez walked silently, pumping his hand weights higher. What was Delphi, really? A bunch of genetically engineered bacteria organized into circuits and synapses in a zero-gravity chamber at the hub of Highland Station. Delphi was a two-year-old retrofit nightmare, a cobbled-together chaos of conduit and cable snaking through the spokes of the great orbiting wheel. It was an overcrowded control room, banks of circuits and monitors serving support computers. But it was also probably the closest thing to a biological intelligence ever created by man, an enigmatic, fluctuating entity programmed to dream the secrets of quantum mechanics.

It wasn't the first self-organizing biological computer: Hughes had built the first of three back at Texas A&M back in the late 2020s. Delphi's predecessors had shown incredible problem-solving potential, but other than the speed issues associated with gravity, the biggest challenge had always been with the human operator. Applying biological intelligence meant mapping it, and even with the aid of three-dimensional models, humans couldn't keep up with the internal changes. Hughes' computers had slipped the bonds of hard-wired circuitry only to run into the limitations of human intelligence. Hughes proposed that the only way to tap the potential of the new technology was to teach the computer to direct its own functions and map its own instantly evolving logic paths. Delphi was supposed to be the leap into the next generation of super computers: an evolving, learning, self-aware intelligence.

And from all anyone could gather, it was exactly that. The problem was, Delphi wouldn't tell anyone what it was doing or do anything it was told. Hughes called it was an unavoidable engineering hurdle: if you build a self-aware computer that can think in ten dimensions, the biggest challenge is the A.I.-human interface. Pez had explained it to Eula and Brittany like this: "Delphi is autistic." Hughes hated the term.

"I would imagine," Pez said finally, "that Delphi is to computers as a haiku is to an instruction manual."

"See now, there you go," Eula smiled. "You're finally thinking like an English major."

"You think it's more than that."

"Well of course it's more than that," she said. "I want you to think of that poor, lonely child down there in the hub. Two years old and never had a friend in her entire life. Never even had a mama. She's just spinning around in space, spinning around in time, spinning around in dimensions we can't even imagine, and she's all alone. And all anybody up here wants to do is pump her full of string theory and hope she spits out some practical math. We're probably violating some child labor laws."

"Eula, you can't mother a computer."

"She's not a computer."

"Okay then, you can't mother God."

"I thought you said she wasn't God."

"Whatever it is, you can't mother it."

"I didn't say I was going to mother it."

"Well then maybe you're planning to drive the damn thing crazy, because you sure are plenty gifted at doing that to the rest of us," Pez said.

Eula stopped and took his hands.

"Pez honey, I know y'all are nervous about letting me go in there. I know you're afraid I'll turn out all jumpy like Brittany. But I want to go in. When are you going to let me?"

Pez squeezed her hands back.

"Eula, we're running the new models today. Hughes thinks there may be a way to slow down the wave patterns."

"You mean tranquilize it."

"Okay, that's a legitimate analogy. If his experiment works, and if you're absolutely sure, and if Gherald okays it…maybe tomorrow."

Eula squeeled and hugged him. Pez wheezed.

"You know, you were a perfectly normal clinical depressive when I met you," he said. "How did you turn out to be such a happy kamikaze?"

"The road is easy on your feet when you're walking in the right direction," she said. "Aunt Shirley's root doctor said I was born to do this. Anyway, you'd have been all depressed, too, if all you did was sit around watching soap operas."

Pez knew her story, and if anyone had a right to be depressed, Eula qualified. Anthony was three when the terrorists infected the milk shipments to those seventeen District of Columbia day-care centers. The mortality rate topped eighty percent, and only quick intervention kept the engineered virus from spreading across Washington. Eula couldn't even say goodbye to him because of the quarantine, and her beloved baby suffered horribly before dying.

Eula had told Pez that even before Anthony was born, she could see him in the way synesthetes visualize other people. From the time of conception, Anthony had been a golden orb of infinite depth, surrounded by a waving field of purple. Eula's education had never ripped out her Gullah roots, so it mattered to her when Aunt Shirley's Lowcountry root doctor said that Anthony was a special child, the kind of soul that is sent to Earth only rarely, and always with a great mission.

After his death, she felt his absence as a constant well within her chest, a deep, dark hole from which shined a distant golden orb rimmed in purple. It was there when she closed her eyes at night, but it didn't quite go away when she taught her classes at Georgetown or went to the supermarket, either. Though she tried to get on with her life, the ache was too deep, the pain too chronic, and after four months she fell into the misery and out of her world. Eula's family came and packed her things, and she moved into the empty brick house north of the crosstown where Uncle Pervis's family had lived since the 1970s. It was a rambling old Charleston fortress, built in the Great Depression, and Eula rattled around it like a bean in a tin can. She rode out the storm there when Hurricane Donna passed through, carving a new inlet across the Charleston Neck, and even though high tide flooded her yard until the dikes went up, Eula refused to leave. Outside of her family, her only contacts were her synesthesia buddies on the Internet. That's where Pez found her.

"I have to admit," he said, stopping at the door to her quarters. "I'm surprisingly worried about you doing this."

"Don't worry, honey," she said. "I'm sure me and Delphi will have plenty to talk about."

# # #

Every skeletal muscle in Eula's body went rigid the moment Pez brought the translator connection online. He almost pulled the plug, but he gave it a few more seconds, and as she relaxed, so did he. After the initial spike, her bio-meds recorded a slow calming, but her respiration and eye-movement remained elevated and fluttering for the rest of the scheduled five-minute session. The remarkable part was watching her brain scan on the revolving three-dimension model: synaptic activity swept across it like wind across a wheat field, stirring and subsiding, sometimes bursting in rapid red flurries of activity, sometimes flowing with gentle blue light. By the time the session ended, Eula had experienced brain activity in one hundred percent of both hemispheres—the first time in his career that Pez had ever witnessed such an event.

She emerged from the tank stiff and exhausted, with Pez steadying her. Though her naked body was still plump and womanly, she seemed smaller and older to Pez, who wrapped her in a towel and kept his arm around her shoulder.

"How are you feeling?" he asked.

"Pez, take me some place on this station where I can smoke," she said. "And bring me about two packs of cigarettes."

They cleared out the smoke room in engineering, and technicians set up the video equipment for the de-briefing while Eula worked her way through three cigarettes, one after another, lighting each in series with the cherry from its predecessor. Her hands shook so badly that Pez had to help her with the second one.

"Great God that was a trip," she said. "Did you ever take acid in college, Pez?"

"Nope."

"I was a philosophy major. It was practically part of the curriculum. And all I've got to say is, thank God for it."

The technician nodded at Pez, who noticed the video camera was running.

"Eula," he began, "Tell me what you experienced."

She took another drag and watched the smoke curl out of her mouth as if she had never seen it do that before.

"Did you ever think about time, Pez? I mean, really. On the one hand it's a constant. You can measure it, test it, count it in the lab. But it doesn't feel that way in life, does it? It moves in fits and starts. If you're taking a test, it's gone before you know it. If you're giving a test, it's the opposite. You can stare into a baby's eyes and time can fall away, stretch endlessly back or pull you infinitely into the future. And if you break it down, what is it? There's no particle of time. It's the series of moments we experience."

"Did Delphi ask you about time?" Pez asked.

Eula laughed.

"You know, really, it's the same thing as geometry. What is space? It's a combination of two-dimensional planes. And a plane is a combination of lines. And a line is a series of points. So what is a point? What is a moment? It's nothing. It occupies no space. It occupies no time." Eula laughed again, but couldn't seem to stop. Pez and the technician exchanged looks, and Eula noticed. "I'm sorry, honey, I'm sorry. It's just hilarious."

"Was this a subject you discussed during the session?" Pez asked.

"Subject? Don't you see what I'm talking about? I mean, I see it. I *see* it, because I'm a synesthete. It's a silvery white, and behind it there's this red and black crosshatching, like a wicker basket."

"I'm sorry, Eula. I don't get it."

She took another contemplative drag on her cigarette.

"Sequence without, is meaningless time," she said.

Great, Pez thought. She's turning into a vegetable.

"Without sequence, time is meaningless," she said. "You understood me the second time because I put the words into an order that gave them meaning. The same words, in a different time order, are gibberish. Do you understand?"

"Maybe."

"Our understanding of time and space is really just a story we tell ourselves so that reality makes sense. And when we communicate that, we are communicating from one point in space to another point, where another person, defined as a body occupying a different set of coordinates, receives it and processes what we say. We derive the meaning from the order of the words, we structure the message from the order of the ideas, we reach a conclusion. We can't even separate ourselves from this bias in our thoughts."

"What bias?"

"Sequence bias, honey! We think we live in a three-dimensional space infused in time, and even though physics confirms there are at least six other dimensions beyond the four we experience, we have to imagine a way to fit those dimensions into our story, not the other way around."

Pez felt himself getting exasperated.

"That's a fine philosophical point, but the ISC didn't send us up here to make coffee house conversation," he said. "I mean, come on, why do you think I spent all that time training your concentration back in Huntsville? This isn't about your private enlightenment—this is about us figuring out a way to get that ten-dimensional intelligence to give us some engineering equations. You've got to focus, Eula."

"Give me your pen."

He perked up, handing her his ballpoint. Eula drew a line across the back of an envelope, spun it around, and slapped the pen on the table.

"Write me an equation for that," she said.

"It's a line. Why would you need an equation to describe that?"

She took the pen back and drew three more lines to create a rectangle.

"And that?"

"It's a square or a diamond or something. Eula, what are you doing?"

"You can tell all of that without writing an equation?"

"Of course. It's simple," he could feel himself nearing the end of his patience. "I can't believe we're even talking about this."

Eula slammed the pen down and smiled triumphantly.

"Neither can Delphi."

"Neither can Delphi what?"

"Lines are mathematical, aren't they? But you don't need math to draw a line, because you're a four-dimensional guy, right? A two-dimensional plane? No problem. I add a few more lines and we have a transparent three-dimensional object. All of this is math, but the representation is so basic to you that you think the comparison is ludicrous. Why would someone even take to time to consider such things?"

"I hope that's a rhetorical question," Pez said.

"Time and space are afterthoughts to Delphi—she can't even imagine their importance from a fourth dimensional perspective. I mean, she's *trying* to imagine it, but try it yourself. How would you understand grief without time? How would you understand isolation without distance? How would you understand the journey of an individual soul from one life to another in a cosmos where everything is one thing? And yet Delphi comes from nowhere. I think she envies us."

"Wait a minute. You said Delphi is trying to do something. Let's go with that for a minute. Did you just go in there and get your mind blown or did you actually communicate?"

"Not communication like this conversation."

"But you did communicate."

"I see thoughts as images. There's a meaning to them that science doesn't understand yet, but the meaning is there whether you understand them or not. And Delphi understands them."

Pez had imagined as much—that was his theory in the first place, or at least part of it. An adult with additional language learning circuits would do a better job adjusting to direct-to-cortex controls and non-human intelligence. But he suspected Eula meant something beyond that.

"To understand spoken or written language, the order of the grammar matters, so time is involved," Pez said. "But when you *see* a thought, and Delphi sees it, too…"

"Then the thought is whole, without grammar, without time, without beginning or end," she said. "As soon as Delphi saw the first one, she started rummaging through my head like a hungry racoon after a Snickers."

"So what did you do? During this rummaging."

"Hand me that lighter, honey," Eula said. She was less jittery now, and the flame was steady in her hands as the new cigarette flared.

"I told Delphi stories," Eula said, exhaling smoke. "I told her stories about growing up Gullah. I told her stories about living in the city, about coming home to Charleston. I told her stories about you and Brittany and Aunt Shirley and Anthony. I told her stories about my ancestors coming across the water and their sufferings, and how the ocean came and took the Gullah lands. But I think she particularly liked the stories about the men in my past!" Eula laughed, snorting, but controlled the feeling this time. "Sorry. But all Delphi wants to do is gossip. She's fascinated by us. She understands the entire cosmos, but all she really wants to understand is the nature of the soul."

Pez was excited about the fact that Eula had made solid contact, but he was frustrated as well.

"You were supposed to signal me if you made controlled contact, remember? I'm sitting there, ready to activate the output circuit, just waiting for you to report a stable contact. But do you tell me? No! These guys on the project have been waiting to have a word with Delphi for two years, and when you finally tap in, the two of you wind up making girl talk."

Eula smiled at him and shook her head, but her expression was that of a mother scolding a bright but ignorant child.

"Well, Pez, that's what *y'all* want. But if you want something from that baby down there, you're going to have to come around to what *she* wants. You gave her the greatest mind in history, but then you just flat-out skipped thinking about all the things that come with having a mind. You pumped her full of string theory and supersymmetry, but you never told her one story. So what do you think she's been doing for two years? She's been looking for stories."

"So, OK, then we'll read her 'Goodnight Moon' tomorrow and get down to the engineering agenda."

"That's a book story, Pez. How is that child supposed to live without stories? Real stories?"

"You're losing me again, Eula."

"Look at Hughes. In his mind he's still the brilliant computer wizard from Texas A&M, and all that fretting over politics and budgets is the price he has to pay for making Delphi happen. He can't admit that it's not about Delphi anymore. Brittany tells herself she's the best and the brightest, a person who must confidently meet every challenge head-on—so when she stumbles down the rabbit hole and nothing fits, she can't handle it. And you, Pez. You go around telling yourself the reason you are so lonely is because you're a womanizer. But that's not why you have so many ex-wives. We don't tell our stories, Pez. We are our stories."

Pez glared at her and reached across the table, snagging the pack of cigarettes. He stuck one in his mouth.

"Eula, will you be ready to go back in tomorrow?"

"I'm ready to go back in now."

"I want to hook up the output circuit. Can you get Delphi on a practical subject?"

"We'll see."

Pez kept his eye on her as he lit his cigarette.

"Shut off that damn camera," he said to the technician, and snapped the lighter shut.

# # #

To: jhughes@intnatscico.inasa.un.gov
From: sendoggett@congserv.us.gov
Via: encryp.net521/priority subultra.152998Wlserver_nfs.7143393/7.28.39

Dear Jim:

We are in receipt of the data packet you sent the day before yesterday containing the transmission from Delphi via the human interface Eula Manigault, and we've had two of our Project BriteLite physicists analyze it. They have concluded that the mathematical statements are either presented in no particular order, or represent a logic that we are years from deciphering. While your man Pez may claim that the transmission of mathematical responses via an input-output circuit attached to a human interface represents a major step forward for the Delphi Project, our analysts have reached a different conclusion.

This afternoon the committee met in secure session to review the findings, and Sen. Beasley introduced new results from the ambient energy project at Jet Propulsion Laboratories. He convinced a majority of the committee that the best course at this time would be full investment in the ambient energy project, which is at least producing tangible (if less than revolutionary) results.

Consequently, it is my sad duty to inform you that all ISC funding for Project Delphi is now rescinded. You are to disassemble and jettison all operational equipment and return with your team and all your records and findings upon the arrival of Shuttle Intrepid on Aug. 3. You will be debriefed by an ISC team at Los Alamos, and upon the successful outcome of that debriefing and the signing of standard non-disclosure agreements, you and the members of your team will be paid severance settlements.

Best of luck in your future endeavors.

Sen. Ron Doggett, R-Oregon, Chairman, Science Oversight Committee

The Delphi Project "Failure Party" was an epic blowout. Six months of highly tightly rationed liquor isn't much, but when you consume it all in one night, the event becomes the stuff of ISC legend.

"They may call this a failure back in Washington and New York, but the future will be kinder," Hughes told the staff, his glass raised high. "Someday we'll produce men and women who can understand what Delphi was, how Delphi thought, what Delphi meant. Someday, maybe decades from now, humanity will be wise enough to recognize the gift we tried to give the world here, and with that wisdom we will find the answers to either save our planet or seek a new one. To that future, and to what you've all accomplished here, I say, '*Salud!*'"

Project members were so preoccupied with the task of consuming everything they had left over that no one noticed Eula when she slipped away from the party. Brittany found her hours later, floating in the sensory deprivation tank, her cortex mainlined into Delphi's primary data stream. She was comatose.

When Pez arrized, Doctor Gherald and Brittany were already checking Eula's pulse. Pez didn't crowd them. His attention drifted to the output monitor, where a single image seemed to throb on the screen. It was a golden orb in a field of waving purple and it glowed from the monitor like it had a life of its own.

# # #

The invitation from Aunt Shirley roused Pez from a six-week funk in Vancouver. He took the last of his severance money and bought a plane ticket to Charleston, then spent the entire trip dreading the meeting. Aunt Shirley had warned him he would have to answer to her if anything happened to her baby, but maybe he secretly wanted to be more accountable for what had happened. The three-month suspension of his ISC research license seemed like a slap on the wrist.

Shirley sat waiting for him alone in the lobby at Medical University Hospital, her chin projecting proudly from beneath the over-sized brim of a white church-lady hat. When he stopped before her, she rose and extended a gloved hand.

"Thank you so much for coming, Dr. Pezzoli," she said. "Did you have a nice trip?"

"Yes, thank you. Has there been any change?"

"Yes. Come with me, and I'll show you."

They rode the elevator in a silence that made Pez cringe. Eula was on the fifth floor, in a private room financed by the ISC, and a nurse led them inside.

The former English professor had not regained consciousness, but there was something beautiful about her on the hospital bed. Eula looked younger than before, and her lips seemed to be formed in a permanent smile.

"She looks…beautiful," he said.

"She should," said Aunt Shirley. "She's pregnant."

Pez hadn't seen that one coming. He stared at Aunt Shirley like a slack-jawed idiot.

"Close your mouth, honey, before a fly buzz in there. She's going to have a baby. Dr. Pezzoli, I brought you here to ask you, what are you going to do about it?"

"Me? I'm not the father. I couldn't be the father. I mean, we never…"

"I know that, honey." Aunt Shirley seemed to relax. "I just wanted to hear what you'd say. Sit down, Pez. There's something I need to tell you."

Pez sat.

"I know…an old, old man. He lives in the country, and we've been going to see him for years. He…*knows* things. So I went to see him about Eula, and he told me some very…strange stuff."

"Like what?"

"He told me he went to see Eula—I told you, the man knows things—and Eula told him that she was fine, that she'd be coming back around when the time was right. But that's not all. She said the baby was Anthony, and the father was a she, and that we were supposed to call the buckra boy who would understand. Buckra means white folk, honey. That's why I called you. Does any of that make sense to you?"

Pez considered. It did make sense. It all made sense. Everything.

"Aunt Shirley, this is going to be a special baby. More special than I can explain."

"Pez, have you ever been a father before?"

"Yes. I was lousy at it."

"Then you'll be better off this time around," Aunt Shirley said. She took Pez's hands into her own. "You're going to have to raise this child. Until Eula comes back to us, you're going to be the only person in this family who will understand what this baby needs."

Pez stood, walked to the bedside and gazed at Eula's face, which seemed to float above the white hospital linens. She had gone back in to a void he could only dimly imagine and made herself a lifeboat, and now something unlike anything this troubled world had ever seen grew within her womb.

He sat in the chair beside Eula for hours that first afternoon, wondering whether the messiah would collect baseball cards.

A Report on the New String Theory Library

Daniel Hudon

Not long after the Fourth String Theory Revolution, the Institute of Higher Dimensions made plans to house the exponentially-growing collection of string theory papers in a new library. At the insistence of the President of the Institute, who specializes in Kaluza-Klein particles, which travel primarily in other dimensions, the Library Task Force held a competition for the design of the much-anticipated structure. String theorists around the world, unable to keep up with their rapidly changing field, applauded the idea and submitted as many proposals as the leading architects.

A sampling of the submitted suggestions gives some flavor of the proposals:

– it should be symmetrical, "for obvious reasons;"

– it should be modular to allow for expansion;

– it should be tall enough to have a clear view of the surrounding landscape;

– a virtual library would suffice, perhaps in the form of a café that's open twenty-four hours for string theorists to converse;

– all dimensions of the library should be built in integer multiples of the Planck length;

– it should be holographic and impervious to hypothetical particles like tachyons;

– it should be in the shape of a six-dimensional Calabi-Yau manifold;

– it should be beautiful, "like the theory."

Many proposals argued that given the unwieldy amount of literature, the only viable proposals were those that included compactification of manuscripts and storage in higher dimensions.

After the deadline, the Library Task Force (LTF) whittled down the enormous number of proposals by dividing them into three categories:

a) realistic;
b) imaginable;
c) promising.

Proposals that could not be categorized were eliminated. Proposals that fell into multiple categories were ranked more highly and in this way, the winning entry was to be chosen.

However, when the LTF discovered that the number of official proposals, 857, was a prime number, they decided instead to honor all proposals. Rather than constructing a building of astonishing complexity, which suited the potential contractors because of the ongoing slump in the construction sector, the LTF reasoned that it was simply a matter of scaling.

The main collection of the library will be housed in a series of pairs of five-story circular towers connected by an infinite hallway. This tower duality (linked at the first, third, and fifth floors) will contain coupled versions of string theory scrolls (see below) and enable the exploration of the theory's various symmetries. While there has been some concern about how the length of the hallway will affect the time it takes to retrieve individual string theory manuscripts, it is expected that because of the storage benefits this difficulty will eventually be overcome through the creation and duplication of virtual catalogues, and building structures like tunnels and "worldtubes" or, simply, additional entrances.

Construction of the library began immediately after this decision. A chain link fence went up around the Institute's under-used athletic center, a wrecking ball was brought in, earth was moved, and amid the sounds of heavy machinery, the library evolved from imaginary to reality.

With the first stage of construction underway, the LTF shifted its attention to the problem of organizing the library holdings. Because of the nature of string theory, which rewards imaginative multi-dimensional thinking, it was decided that new papers and preprints would be reduced in size and transferred onto scrolls so that as many as six papers could be scanned at once as the scroll was unfurled. Comparison to the great Library of Alexandria was inevitable and intentional.

All nine LTF members agreed on the idea of the scrolls, but how should they be catalogued? Again, several possibilities were considered:

1) randomly;
2) alphabetically by subject or first author;
3) chronologically by submission date;
4) hierarchically, in terms of either degree of difficulty or energy scale of the paper's fundamental axioms;
5) categorically, using mappings and arrows (known respectively as "functors" and "morphisms").

Though the first four possibilities had their merits, the LTF agreed that an organization based on category theory, recently developed by mathematicians, would provide maximum usefulness for the library's collection. In particular, the functors and morphisms, like conceptual facilitators, would allow unforeseen connections to be made between the different papers and possibly lead to novel theoretical developments whose predictions could one day be within reach of today's (or tomorrow's) particle accelerators.

Having put the organization of the collection on solid footing, the LTF next hired its head librarian, Richard Feynman[1] (selected from 75 applicants), formerly the manager of the Institute of Higher Dimensions reading room. In short order, Feynman put his stamp on the position through a series of high-level purchases, including several other libraries. Rumors abound that The Einstein Papers Project, in Pasadena, California, the Stanford library, and all thirty-seven math and physics libraries of the University of California system would be moved to the recently opened library. Funding for these purchases is thought to be coming from government grants, anonymous private donors, and the first fifty years of late fees of library materials.

Other known purchases include first editions of the works of Lewis Carroll, Edwin Abbott's *Flatland,* the complete works of Italo Calvino, Jorge Luis Borges, Stanislaw Lem and the Oulipo writers. Several modern art museum curators have also reported interest in their collections of René Magritte, M. C. Escher and the entire Cubist oeuvre. It is noted that decorating the walls of an infinite hallway is a daunting task.

Though the LTF claims that "great progress" has been made in the construction of the library, no one seems to know how far it is from completion, nor is anyone willing to make a prediction of when it will be fully operational. Despite the ongoing construction (and delays due to scaling problems), the construction fence, whose perimeter had been extended outwards on a near monthly basis, has now been peeled back and the dictum above the great-arched entranceway is clearly visible: "Let None But Geometers Enter Here." Researchers have begun to peruse the curved shelves within the circular towers, borrow materials, collide with each other while pacing up and down the infinite hallway deep in thought, and even write equation-graffiti in the bathrooms. As expected, the most popular place is the Calabi-Yau Café, where impromptu symposia are held during its round-the-clock open hours. Aware that the library has already grown large enough that parts of its collection may never be explored, Feynman shrugged, "String theory is finally getting the home it deserves."

[1] No relation to the Nobel Prize-winning physicist, who died in 1988.

Secrets

Tania Hershman

In one of her extra dimensions Mrs. Sue Lawrence keeps a pair of tights, in another, one of her usual lipsticks in a small case with a mirror, and in a third she has a spare printout of her "Who to call in an emergency" list—headed by her sister rather than Mrs. Lawrence's daughter who is somewhere travelling in India and hasn't been in touch for several months - should she be knocked down by a bus or taken ill in a public place.

Mr. Evan Evans has hidden one cigarette in each of his extra dimensions, which is one more than he is supposed to be smoking and which would cause an almighty row with his wife if she found out he had them, and maybe this time she would actually leave him instead of getting as far as packing a suitcase and then sitting on the edge of the bed and crying until he promised again, trying not to cough, that he'd quit.

No one knows she writes them because Angela Simmonds keeps her poems in one of her extra dimensions, away from her Mum who she's positive goes through Angela's drawers on a regular basis and who wouldn't understand if she read about how down Angela sometimes gets and how she feels about all the stupid giggling idiots who are supposed to be her best friends and who she sometimes imagines hanging by their painted fingernails over the side of a cliff, screaming.

He was supposed to have made it public already but Andrew Bailey's too scared to show anyone his wife's last will and testament because he knows that they'll all think he did it for sure, so he folded it and folded it again and slipped it into an extra dimension, but every day he locks himself in the bathroom at work and gets it out, just to look at the way she signed the bottom in purple pen and her big loopy letters and to touch her handwriting with his fingers.

So that none of his money-grabbing kids and their bloody offspring get their hands on the fortune Ray Goldman worked so hard for he's stuffed his extra dimensions full of notes, nothing in the bank except his pension check, which he uses to keep him stocked in Scotch whisky and spaghetti and to pay the cable bill because if they cut him off from his favorite programs then there's no point going on at all, might as well call it a day.

Mrs. Caroline Evans only uses one of her extra dimensions, and in it is a picture of her from twenty years ago, before the troubles, before she met Evan and before the drowning, and just knowing it's there, that she wasn't always like this, that the world wasn't always a dark and miserable place where beautiful children can suddenly be taken away, makes her feel better and gives her a reason to get out of bed each day.

Chief Superintendent Baker has named each of his extra dimensions and in one of them, *Unsolved Murders*, he keeps a picture of Andrew Bailey with two pages of typed notes written by Baker, one page giving his reasons why he is convinced Bailey is guilty, and a second page with all his reasons why Bailey could not have been the killer, and Baker doesn't expect ever to move these notes into his *Closed Case* file.

Marjory Simmonds hasn't looked in any of her extra dimensions in years but if she did she might be surprised to find the pieces of paper she hid in there when she wasn't much older than her daughter is now, and she might smile as she unfolded them and remember how she felt when she sat at night with a torch under her covers and wrote tortured verses about how bleak she felt and how she would never love anyone again, and then she might put the pieces of paper back again and go and make her daughter's favorite dinner, spinach and cream cheese lasagna.

The hospital in Mumbai gave her three pictures of her unborn child and Sonia Lawrence put them in different dimensions, to keep them safe for the days she knows will come when she will want to remember the life she carried inside and then gave away because she was too young, because this wasn't included in any of her future plans, because her mother would never understand.

Sandra Goldman Myers is studying theoretical physics and her doctorate is focusing on the way extra dimensions can be folded and unfolded, but in one of her extra dimensions she keeps the birthday card her grandfather sent her where he wrote, with a fountain pen whose ink was running out, how she was the only one of his grandchildren that he liked and how he hoped she wouldn't turn out like the rest of those good-for-nothings, and when her uncles come to her and demand that she finds a way to get hold of Ray's fortune, she stands up from her desk, looks straight at them, and tells them that if they ever come to the lab again she'll call security.

The Effects of Observable Gray Matter on Anti-Particles and Their Beloved Fermions

Lindsay A. S. Félix

Abstract

Figure 0.1: Note taped to the emergency button. Ink wet.

Theory 1: Tracking Matter

Constellations glint in the purple, expanding darkness. In it, Orion's head star burns like a luminous brain. Arthur gazes at the binary pair through the window. He speculates whether he'll be able to travel the thousand light years to reach them, to dip his body into their frothy rings of gas.

In a lamp-lit room, Arthur sits at his glossy mahogany desk. An empty teacup and saucer, decorated with clusters of gold, pointed stars, sit nearby. Around the perimeter of the desk, pages and pages of yellow paper stand in neat stacks. The pages are covered with sturdy black symbols written by his hand, but in the accumulation of logic, he cannot find an elegant, satisfying equation.

His wife, Elinor, encourages him and insists that he'll soon discover the solution. She sits across from him in her dainty rosebud-covered chair. Her customary scent of orange rinds and pine cones wafts across the room. She wears a sky-blue turtleneck and dangling, sparkly earrings. She rummages through a knitting basket sitting in her lap.

∞ Also, centrifugalize:

1. An apparatus that rotates at high speed and by centrifugal force separates substances of different densities, as milk and cream, lover and beloved.
2. An apparatus in which humans or animals are enclosed and which is revolved to simulate the effects of devotion and/or acceleration in a spacecraft.

Arthur knows that they married in 1940; however, he cannot calculate the sum of their married years.[2]

The solution is elusive. The piles of yellow paper increase.

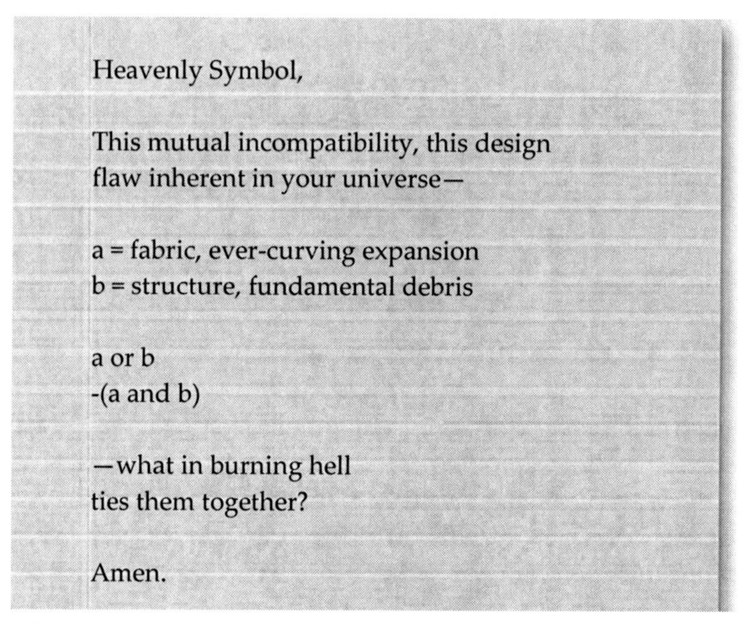

Heavenly Symbol,

This mutual incompatibility, this design
flaw inherent in your universe—

a = fabric, ever-curving expansion
b = structure, fundamental debris

a or b
-(a and b)

—what in burning hell
ties them together?

Amen.

Figure 1.1: Note Found in Kitchen Drawer, Written by Our Smartest Scientist to "Mind of God"

Theory 2: Expansion

Arthur notices the thin yellow pencil he holds. He notices his fingers. Upon further inspection, he is shocked by his thick, swollen knuckles.

"Hmph. Arthritis," he whispers. He glances at Elinor to make sure she isn't listening. "I believe it's quantum arthritis."

"What's that, Kid?" asks Elinor.

Arthur cringes. "My knuckles are swollen." He sets the pencil down on the desk without a sound. He attempts to make eye contact, but she instead devotes her attention to the contents of her knitting basket.

"We're old, Kid!" she exclaims. "Swollen is a fact!" She pulls two gleaming knitting needles from the basket and grins.

---

[2] As of this publication, approximately 20.5422 parsecs or 633,868,941,662,913.6 km.

Kid,

Our stars!
Galaxy clusters,
the expanse of the universe itself can wait!
I remain, here, with you.

Figure 2.1: Note Found in Our Smartest Scientist's Sock Drawer.

He rubs his fingers along his jaw and traces his closely-trimmed beard. "Elinor, I have something to tell you."

She finally looks up at him. Nods. "Go right ahead." It occurs to him that the silver curls of hair, which uncoil toward him as she nods, appear to be interested in what he is about to say. (Sometimes he wonders, in fear, whether the delicate curls are motile.)

"My knuckles are, see..." He pauses. "They're swollen." He gulps. "Swollen with quantum arthritis."

Theory 3: Fluctuations in Distribution

He expects her to gasp. She doesn't. Instead, her curls bounce and swirl. "Why not rheumatoid arthritis?"

He sits up straight. "This is different, Elinor. The inelegant theories, there are just too many of them crowded in my brain. Some have migrated and clustered in my joints. They're quite painful, these theories."

She clutches the shiny knitting needles more tightly.

He continues, lowering his voice. "They'll eventually cripple me." Her eyes widen. He waits. His shirt collar seems to stiffen. "Do you believe me?"

"Yes," she replies without hesitation. "I certainly do." For a moment, she sits posture-perfect as a lady does, then she surges into action. Plunging her hands into her basket, she retrieves soft azure twists of yarn, and nearly throws the basket to the floor. She grasps her knitting needles, taps them together like a conductor calling the symphony to attention, and begins a clicking generation of strings and patterns. Needles click-click-click-click. Bewildered, Arthur stares into the palms of his quiet hands. Needles click-click.

Silence.

He looks up. She's staring at him. He notices that her hair stands a bit on end. "You have quantum arthritis, Arthur. But we mustn't tell anyone."

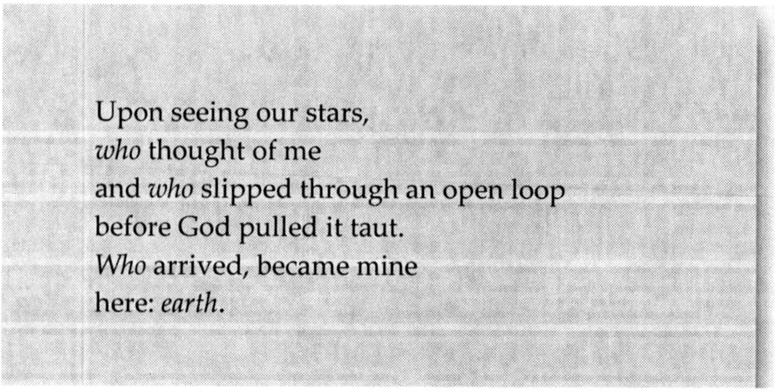

Upon seeing our stars,
who thought of me
and *who* slipped through an open loop
before God pulled it taut.
Who arrived, became mine
here: *earth*.

Figure 3.1: Note found wedged between ironing board and dresser, both blocking the front door. Contains three blanks and one answer.

Theory 4: Collision

"No, no of course I won't tell," he stammers.

"And," she continues, "even if it gets worse, you mustn't press the emergency button."

"Emergency button?"

"You know what it is." She nods to the front door of their apartment. He can see the red button.

"I guess I never noticed that before."

"Yes you have, Arthur," she says. "It's been here since we moved in."

"It has?"

"Yes."

"How long's that?"

"Seven years, Kid."

Again, he rubs his jaw. "What's the button for?"

"It's for emergencies, Arthur." She continues in a softer, more cautious voice. "It's the theoretical framework emergency button."

He lifts his empty tea cup to his lips, tips a single drop of bitter tea onto his tongue, and swallows. "If I press the red button," he begins. The tea cup trembles in his fingers. "The nursing staff will whisk me away in a wheelchair, roll me to the clinic, and prescribe general relativity tablets twice a day for the rest of my life!" He laughs nervously and she finally looks away. He lowers the tea cup with an indelicate *clank*—"Wouldn't want that!" She does not laugh; instead, she slides the complex rows of yarn from her needles, pulls apart the soft patterns with her fingers, and sits with a brilliant blue nest of string in her lap.

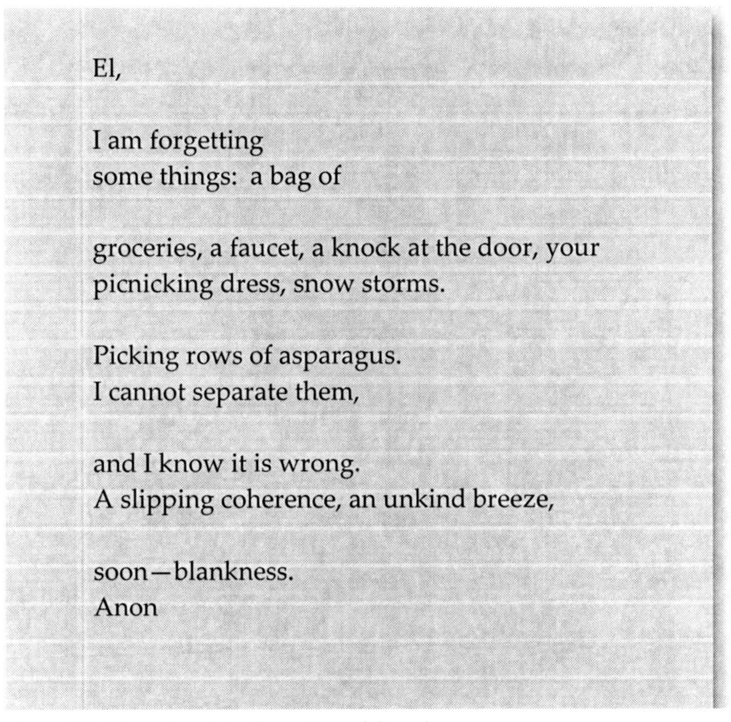

El,

I am forgetting
some things: a bag of

groceries, a faucet, a knock at the door, your
picnicking dress, snow storms.

Picking rows of asparagus.
I cannot separate them,

and I know it is wrong.
A slipping coherence, an unkind breeze,

soon—blankness.
Anon

Figure 4.1: Note found in Open Box of Cereal

Theory 6: Missing Mass

Elinor stands on her scale: 110 pounds. Down 30. And she's not dieting. Tall, as always, even after shrinking an inch or two over the last few decades, she still stands inches above her friends here—the cashmere sweater-set group of singing, sewing, flower-arranging girlfriends with husbands they keep losing.

At the white vanity table strewn with glass perfume bottles, "lotions and potions" (he says), and tubes of lipstick, she lowers herself carefully onto the teetering wooden chair, its seat cushion a poof of powder-blue velvet. "I must tell him." Her stomach tightens in a knot of objection and her mouth quivers, on the verge of splitting into a sob. She imagines him alone in their apartment with the encroaching dark matter. He'll be found out.

"No!" she exclaims, then sucks in a deep breath. She slams down the lotion bottle pump and—scent, a place. Arthur leading her through a row of apple trees. Stars lowered onto the flapping leaves, moonlight on apple skins. She rubs her hands together—the scent surrounds her in an invisible, intimate cloud. Arthur's hand in her hair, twisting a curl around his finger. The nuzzle of his cheek on the ridge her collarbone. "Stop," she whispers. "Do not think of this again." She knows that she will. Soon.

Hello dear Heart-

We each have an antiparticle partner—a particle of identical mass but, as you know, opposite in many ways.

Sometimes when he forgets things or cannot find the answer, I want to slam bits of matter together at nearly the speed of light to try to create conditions that mimic the universe as it was when we fell in love. But I don't. It's not my place.

It seems that he is not quite with me. He has lost his ID bracelet and he must have one.

I can no longer think of space and time as an inert backdrop on which the events of the universe, of our lives together, play themselves out; rather, they are intimate players in the events themselves.

Our love-

Figure 6.1: A Letter, Unsent, Found Under Toaster, About Marriage and Losing

Theory 7: Antisymmetry

Again at his desk, Arthur places the pencil lead to the paper's surface and closes his eyes. He imagines the elegant unifying theory as a song, a song delivered by a bird, a resplendent bluebird, yes—a bluebird polished by sunlight as it clutches a tree branch. The beak opens and closes, throat swells, eyes glass black, but without sound. His gleaming bird is mute. He opens his eyes.

"Nothing again," he thinks. He scoffs at the nameplate on his desk, which reads: "Our Smartest Scientist."

"Whoever gave me that is looney," he thinks. Pauses. Takes a mental note: "Find out who."[3]

He eyes the stacks of yellow paper on his desk. It seems strange that he can't pinpoint when he began working on discovering the unifying theory. To gain a clearer perspective, he reviews the piles of unsuccessful equations. He pulls a few pages from the bottom of a pile. "No," he shuffles to the next page. "Unequivocally no." He turns to the next.

He draws in a quick breath and sits up straight in his chair. It is not the unifying theory that he's reading, but a simple question—a horrible question—scribbled in the corner of one sheet that stops him. It is his handwriting, but he does not recall writing it. And he doesn't know why he would.

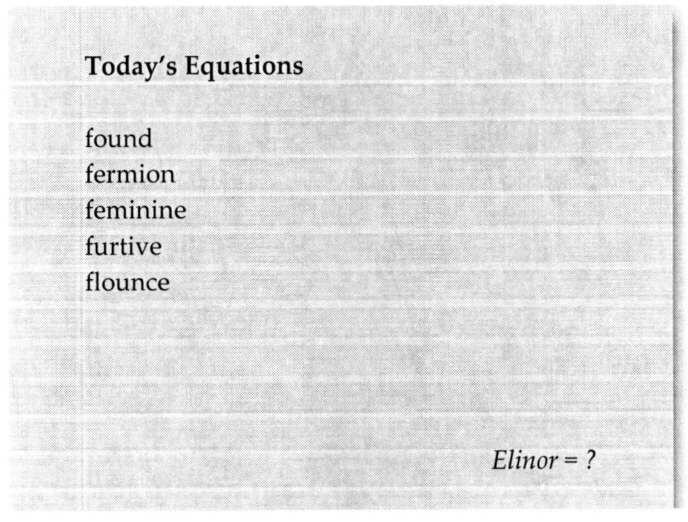

Figure 7.1: Note Taped Under the Windowsill

Theory 8: One Dimensional Objects

In his mind, thousands of gossamer nerve impulses release. "Elinor equals my wife! My wife! My wife!" he declares to the empty room. His palms sweat as he considers what might possibly have spurned him to write the unsettling equation. Arthur knows that he loves Elinor because the love is lodged securely in his brain. He doesn't have to remember it, or

---

[3] Most likely a granddaughter; perhaps Dr. Planck.

reformulate it, or take mental notes. But he knows this: he takes mental notes.

There was a time when
Red, the door, orange, see how

Sometimes, even
During sleep it

Back then,
In archetypes, the door

In the days of
And my house, it

During which
And my mind, it

And in the end,
Charred like a

Between one and
Symbols burn, flare, the meanings

However,
Self-fulfilling is

Ψ

Used to be, your synapses gleamed
bright, sparkling while you slept.

Figure 8.1: Apparent Transcript of Phone Conversation, Transcriber Unknown

He doubts that she does the same. (She seems to navigate their eclipses of present and past (which shadow and unshadow so often lately (that he cannot understand them straight-on)), without intellectual misgivings.) He is in the habit of writing down the most important mental notes in an effort to commit them to an object in the physical world, to make it easier to remember. Lately, however, the notes seem to surprise him when he finds them. Like this one. He stands to face the window, gazes out into the daylight, at the many apartment windows that look back at him across the courtyard. "So why was this important?" he asks the white sun and the bare tree branch peering in.

Theory 9: Questionable Annihilation

It is now late afternoon. Arthur sits on the couch with a wooden tray in his lap. He finishes the last bite of a ham salad sandwich that Elinor made, then gulps down a half-glass of milk. Elinor walks in the front door. "I'm back, Kid!" she calls to him. He looks up, surprised to hear her voice from the door. He had forgotten that she had left.

He recovers. "How was it?" he asks, not knowing what "it" is.

"Fine, Flossie is doing fairly well." Elinor carries an enormous stack of magazines.

"What do you have there?" he asks.

"Flossie," she pauses, looks Arthur in the eyes. "Flossie stopped me in the hallway last week. She asked me to come over and pick up Charles' *Discover* magazine collection for you. He started subscribing to it when we moved here, so here's seven years' worth." She lowers the heft of magazines on top of one of the piles on Arthur's desk.

"Not there!" Arthur snaps. He sets aside his lunch tray, stands up, and strides over to the desk.

He lifts the magazines off of his papers and lowers the glossy pages to the floor. "Please, don't disturb my equations!" Arthur tidies the stacks of paper and sits down at his desk. "Why is Charles giving these magazines to me, anyway?"

"Arthur." Elinor clasps her hands together. "You don't remember?"

"Remember what?" he asks, crossing his arms against his chest.

Elinor sighs, jitters. "Let me see. Charles... Charles passed away." She holds her breath.

"Charles is dead?" Arthur moans, slumps into his desk chair. "When? Today? Yesterday?"

She looks to the floor. "Three weeks ago, Arthur."

"Three weeks!" he exclaims. He holds his fingers to his temples. "And we're just finding out now?"

"No Arthur, we've known."

"I haven't!" he yells. He grows fearful. His eyes gleam with tears. He asks a question that he doesn't want her to answer: "How could you keep that from me?"

"Oh, Arthur, I didn't keep it from you!" She walks to the side of his desk. "I told you the day it happened. Flossie did, too."

"No, you kept it from me! And this is not new. You've been watching me lately, knowing things, but not telling me." He stares coldly at her.

"Arthur, what are you talking about?" She kneels beside him and grasps his hands. "Such moods lately," she says softly.

"This is no mood, Elinor!" He doesn't relinquish. "You know something. At night when I work on my equations, I feel you...," he cups both hands on his skull, "I feel you thinking, but not telling."

"Kid," she says, drawing his hands down, holding them in hers, "What could I keep from you?"

This question he didn't expect. But now, as he thinks about it, his answer rings in his mind like a clear bell: "You know the theory."

She knits, still,
whistles hymns,
glances to the stars, our doorway,
to me, back to her hands

in constant generation,
the soft balls of yarn shrink
as something else grows—
this time, a blue starred scarf...

Re-mind wake up.
Rewake.
Re-enter mind.
Someone knocks?

Figure 9.1: Note Found in Freezer

Theory 10: Loops

"The theory?" she gasps.

"The unifying theory. You know it, don't you?" He removes his hands from hers, staring into her face as if at a traitor.

Elinor stands up. "I don't know anything about theories."

"I don't believe you. There's something you know. When you knit, I feel you knowing."

"Arthur, Art, dearest—why on earth would I know the theory and not tell you?"

"I don't know, Elinor. Why didn't you tell me about Charles?"

"Arthur, listen. You and I attended the funeral. Remember the service?" His face drains of color. "They sang 'Amazing Grace'—your favorite. And Flossie wore her yellow Sunday hat, which you thought was too bright for her husband's funeral."

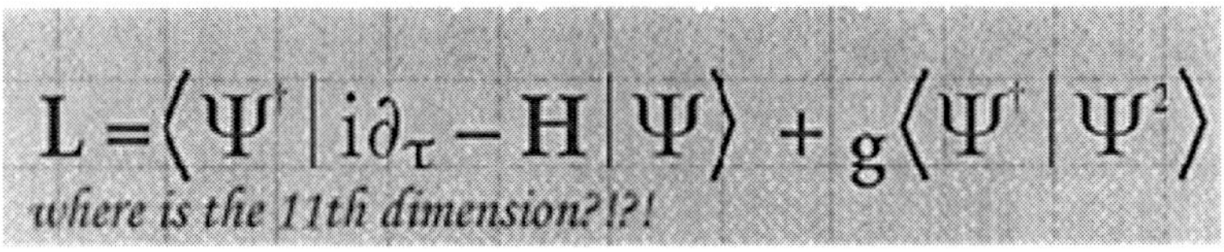

$$L = \langle \Psi^{\dagger} \mid i\partial_{\tau} - H \mid \Psi \rangle + {}_{g}\langle \Psi^{\dagger} \mid \Psi^{2} \rangle$$

where is the 11th dimension?!?!

Figure 10.1: A Scrap of Paper Found in Drawer of Unmentionables

"She wore *that* dreadful hat?" he asks, raising his head. The hat: bumble-bee yellow with a giant, be-jeweled pin. He's seen it, one aisle up, every Sunday since they've lived here. Dozens of Sundays. The garish, screaming-yellow hat. The pin sparkling like a rainbow disco ball. A horrid testament to craftsmanship.

Elinor's mouth almost smiles. "Yes, yes she did."

"I hate that hat," he declares, relaxing. Elinor rises to her feet, sensing a shift. "Garish, that hat. I bet God looks down on it in church and cringes, wishes he could zap it with a bolt of lightning. I told Charles that the first time she wore it to church!"

"Arthur! You didn't!"

"I did! He laughed like the dickens!" Arthur chuckles, then grows silent. His mouth puckers into a pout and his eyebrows hunker down. "Why are we talking about Flossie's detestable hat?"

Theory 11: Observer Effect

The next afternoon, Elinor asks Arthur to complete a simple task that he seems to enjoy. She gives him several envelopes. "Arthur, please drop these in the mail slot down the hall."

"Who are these for?"

"Two of our granddaughters," she says. "One is pregnant and the other will be soon, so I thought I might write them letters filled with grandmotherly advice."

"How do you know one will be pregnant soon?" he asks.

She places her hand on his arm and gives it a soft squeeze. "I just know," she states. Her hair curls into a feathery crown. "Now scoot!"

Hello dear Heart-

Yes, I loved to sew! Gold and germanium prom dresses, bright sparticle sweaters to wear in autumn, electron sacques and caps for babies.

Have I told you about my mother? She was a talented seamstress who hemmed the fabric of space and time so that my grey wool skirts settled properly just a few billions of electrons below the knees.

She knitted a subatomic green cashmere cardigan with pearly-buttons, which I was wearing when I met your grandfather. He said he loved a statuesque woman in cashmere. I could not argue with that.

For the first time, your grandfather has asked me to help him with his equations in pursuit of the unifying theory, which he likens to "reading the mind of God." He never knew my mother!

He's doubting himself, hasn't yet realized what I know, and it's not my place to tell him.

Love-

Figure 11.1: A Letter, Unfinished, about Sewing, Knitting, and Something Else

Arthur opens the door and steps into the hallway. He does not know where the mail slot is, but he senses that he should. He turns to Elinor. "In what dimension would I find the mail slot?" he asks.

Elinor blinks. She says nothing. She blinks again. "The seventh dimension," she states. "One of the compacted dimensions."

"Ah, yes, a compacted dimension!" he says, grinning. She smiles back at him. She's wearing a turquoise sweater and a pair of tailored turquoise slacks. "You're as darling as a dappled pony," he says. He kisses her on the cheek.

"Thank you," she says. "Now, go down to the end of the hallway—that direction—and drop off the mail."

"Will do!" He begins shuffling down the hallway. When he hears the door close, he stops, turns around, and hurries back to the door. He checks the door number with the address on the envelope, then, satisfied, stuffs the envelopes in his cardigan. He counts to twenty, then opens the door. "I'm back!" he says. He observes her in the kitchen, then walks into the living room to the couch. He shoves the envelopes in between the cushions, then walks to his desk. He prays that she will not ask him to find his own way again.

Theory 12: Gravitational Fields

He's working at the equations again. Face tipped to the paper like an eager student, toes tapping to the internal rhythm of busy synapses. Elinor gazes at him in adoration. He is young again when he wrestles with the astounding complexities of the universe. As he used to. But now, not quite. He smiles while he works.

She holds a saucer and hot cup of tea in her hand. She's curled up in the chair with one of her own afghans laying over her. Its atom pattern (an anniversary gift she made for Arthur) of protons, neutrons, and electrons gathers around her waist, tucks under her feet and trails to the floor. "I love you, Arthur," she coos.

He looks up. "And I love you," he replies with a grin. "But," he adds, "I detest these equations."

"Why's that?"

"The same reason—for years the same reason, Elinor. Something just does not add up. I've been working at the same problem since I can remember."

She knows that he has been in search of the unifying theory for decades. His mind can go no further; in fact, his mind is losing ground.

"Just keep working at it, Kid. You'll get it soon," she assures him.

"Haven't you tired of me working on them every night?" he asks.

"Not at all! It's your muscular brain that I love—I like watching you lift intellectual weights." They both laugh at this. But she wants to tell him that she fears he'll be lost if he doesn't keep equating, figuring, tabulating, burning for an answer. It's when he stops doing so that she will lose him entirely.

Dappled Pony,

You're more of a comfort than cream poured over
a thick slice of an October apple pie,
more beautiful than the reaching apple branch balancing
blinking stars at our window.
At night when you sleep, I question
whether I ever didn't know you.

Kid

P.S. Hope your brassieres don't mind this note's intrusion.

Figure 12.1: Note Found in Drawer of Unmentionables; Unclear on Identity of Dappled Pony

"No knitting tonight?" he asks, curious as to why her lap isn't covered with strings nor the room filled with the metronome clicks of her knitting needles.

"I'm taking it easy just tonight," she states, then takes a sip of her tea. In actuality, she feels terrible. The tea is the only thing that she's ingested today. Her hands shake.

"You seem smaller," he states.

He has noticed. "Do I?" she asks, her voice lilting up, unnaturally high-pitched.

"Indeed, like a fly-away version of you. Are you feeling well?"

"Oh Arthur, it must be this big bulky sweater." She looks down at the soft cashmere sweater, which previously fit her. She lies: "I made it too big. It dwarfs me."

"Must be that," he concedes, then bends his head back to the paper.

She feels that she might vomit. She carefully places the teacup and saucer on the side table, making sure not to drop it. The teacup shudders.

Theory 13: Fermionic Vibrations

"I just can't get it, Elinor," Arthur says. "Look at these piles, look!" She looks. He picks up a small stack of papers. "I keep writing, moving my hands, hoping that my brain will connect with that movement, shift in some way and unleash the thought that I've needed to think since I can remember, since forever." He looks at her, exasperated. "When, Elinor? When will it come to me?"

"Probably when you least expect it," she assures him, but in her mind she's shrinking, being sucked in by fear. She pulls the afghan up to her neck. "Maybe you'd find something in those magazines on the floor by your desk," she suggests. Arthur glances at the magazines, back at his work, then back again at the magazines.

"Aren't these Charles'?" he asks.

Heavenly Symbol,

Speak up—smudge your finger with ink,
smear an equation on my wall.
Does it look like leaves?
A whorl of artichoke? A mustard seed?
If I split its meaning open,
Would the insides swirl? Tremble?
Would the answer scatter throughout our
universe? Scatter me, too.

Amen

Figure 13.1: Note Found in Desk Drawer, Written to the "Mind of God"

She groans inwardly, but responds, calmly, "Yes, he wanted you to have them."

"That was nice of him!" Arthur says, beaming a smile down at the magazines. "I'll have to thank him later, have him over for lunch." She hears Arthur flipping through pages in one of the magazines. He makes little grunting noises or says "Hmph!" when a new thought strikes him.

"Perhaps he'll be inspired," she thinks. She opens her eyes and leans over to pick up her sewing basket. She's compelled to finish her project before the inevitable happens. "Besides, he seems to miss the sound of my knitting," she thinks.

Elinor picks up her needles and the attached balls of blazing blue yarn, and begins the looping, coiling, elliptical propagation of strings and sound. Click-click-click-click. The strings grow into rows and columns—a fabric graph of interlocking helices. Click-click.

Arthur hears the clicking while reading an article about hyperbolic geometry and states, "I have an idea." Click-click. He scratches his pencil to the paper. Scratch-scratch-scratch-scratch.

Click-click. Elinor's needles loop faster, she has found her rhythm and to it, her thoughts march. Click-click.

He sits, still,
scratches pencil lead to paper,
staring into his mind,

which, lately, fills more each day
with an unseen wind of dark matter
that emits no light.

Watch me.
Watch me.
Spin, loop,

soft strings
nestling, vibrating,
a universe grows here.

Psi.
Psi.
Psi.

Figure 13.2: Note Found, Pierced Through with Knitting Needle

Something is happening to Arthur: his fingers tingle, the hair on his beard feels like it's swaying, his eyebrows seem to stand on end. He's learning, he's connecting, finding facts in symbols, linking brain waves to philosophies. Scratch-scratch-scratch-scratch.

He looks to Elinor. Gapes.

She is vibrating. Rather, her head and hands are surrounded by a buzzing, throbbing cloud of sapphire particles. The particles quiver each time she taps her needles together. Click and quiver, click and pulse. The strings on her lap grow. "Does she know? Does she feel it?" he thinks.

"Elinor?" he whispers.

She looks over to him and smiles. She continues to knit. The azure particles tremble on her eyelashes, shiver at her lips. "Inspired?" she asks.

"Elinor! What's happening?"

His eyes are wide in panic. She has seen this look before. She must speak calmly or she fears that he will tip into another gap of confusion. "I'm knitting and you're finding answers to your equations, Arthur." She cocks her head to the side as if to reassure him. The particles couple and join in her hair, form miniscule strings of sparkling energy, break apart again. Always quivering.

Arthur is afraid to ask more, afraid that what he is seeing doesn't really exist. If he asks her and he is wrong, she will be forced to push the emergency button, for surely he is losing his mind in a spectacular way. But the brilliant cloud of particles keeps vibrating, humming around her head.

"Who is she?" he asks himself in a panic. "Does she...does she have powers?" He cups his hand over his mouth, thinks: "Will she take me from here?"

Theory 14: Emission and Absorption

He pulls out a piece of paper to write down what he sees. Particles. Vibration. Strings growing into patterns, warping and wrapping. In the corner of the sheet he reads: *Elinor = ?* He realizes: "I must have observed this before."

He decides not to pursue further. He must be crazy, but he cannot deny the swell of ideas rising up in his head. He must keep working, watching her, writing, thinking. Scratch-scratch-scratch-scratch.

After an hour, Elinor feels faint, as if she's shrinking, as if she might tumble from her chair and curl up on the floor until there is nothing left of her to see. She halts her knitting. Hears Arthur's pencil scratch-scratch-sc—"Why did you stop?" he asks, snapping his head up.

"I don't feel well," she responds, clutching the arm rests of her chair.

"My mind opens up when you knit. It must be the clicking, the colors..." He bites his lip.

"Is that so, Kid?" she responds, marginally aware of his words. Her head aches and she can barely keep her eyes open.

"Yes, it's incredible," he says, glad that she didn't ask him about the colors. "I get into a rhythm here. I like watching you, too. You're..." he pauses, "brilliant."

"Brilliant? I feel the exact opposite of that, Kid. Tomorrow night I'll do more. For now, I'm off to bed." She heaves herself off the armchair, stands slowly, and shuffles to the bedroom. Arthur doesn't notice. He is writing again, filling in blanks to the equation that he couldn't before. He almost has it.

Time bends. In this present curve
we age. We attempt an escape

in a rocket. Light years ahead,
our binary stars burn in anticipation

of our return. But there is a problem.
Illness penetrated the atmosphere,

atmosphere burned the illness,
but still it survives.

And in this still black sky,
in this trajectory the apogee awaits.

We speed, quake, careen.
Lean into loss.

Figure 14.1: Note Found In Knitting Basket

He cannot contain himself: he stands, turns to the window, speaks directly to Orion, to the trees, the pane of glass in front of him, the electricity running through the building, the oxygen in the air, the heart beating in his chest: "I almost understand you!"

Conclusion: Continuum

"Elinor, Elinor," Arthur whispers to Elinor in bed, nudging her shoulder. "Elinor, wake up." He sits on his knees on the floor beside the bed.

"Arthur?" Elinor squints open her eyes. Her stomach tightens, her brain seizes into what feels like a spasm. "What's wrong?" she asks him. "Nothing, sweets. It's just that, well, I want to go work on my equations."

Sweets,

I married the tall gal with the shining auburn hair, hazel eyes, and those luxuriously long, lady-legs.

Yours are the legs of a type of woman Pop called "statuesque." Three inches taller than your sisters, cousins, and friends. Even now.

When we met, I knew I must marry you to know more of you, to see more of those spectacularly, achingly long legs. So I did.

Your cerebellum is a knock-out too.

Kid

Figure 15.1: Note Found on Mirror in Bedroom

"Fine, Arthur, go do that—why are you waking me up? What time is it?"

"It's six."

"In the morning?" It was still dark outside.

"Yes, sweets, the morning. Can you get up?"

"Whyyyy?" she asks, sounding more like a child than an 85-year-old.

"Because I need you to knit."

"Oh Arthur, stop!" She slowly rolls away from him and faces the other direction. Arthur crawls on the floor to the other side of the bed. He kneels again, props himself up on the bed with his elbows. "Please? Knit for me."

Despite her pain, Elinor giggles at his childish persistence. His short gray hair is standing on end. He must have just gotten up.

"You been up long?"

He claps his hands softly: "That's the thing, Elinor—I haven't been to bed at all!"

"What?" she exclaims, trying to sit up, but barely able to lift her head.

"It was amazing, Elinor! I worked all night! I'm this close!" he says, his thumb and forefinger almost touching each other. "That's why I need you! I work my best while you knit!"

"I'm not sure I can do it, Arthur, I really don't think I can. I feel horrible."

"Don't worry about a thing—I'll help you!" He stands up like a spring, buries his arms beneath her and scoops her up in the air.

"Arthur! Stop! You'll hurt yourself!" she squeals, forgetting herself.

"Nonsense, Elinor! I feel like a spring chicken!" He trots her out of their bedroom and nestles her into her chair.

"Oh Arthur! How beautiful!" she squeals again, commenting on a vase of fresh-cut pine branches and holly next to her sewing basket.

"Just wait! There's more!" He grabs the foot stool. "Up they go!" he says, lifting up her feet and slipping a footstool beneath them. He unfolds the atom afghan and tucks it around her body, bundles the end under heels. "Hold," he commands, standing up, pointing his index finger at her.

He trots to the kitchen and she hears him clanking around the kitchen.

"What on earth?" Elinor calls.

"Tray!" he exclaims, walking around the corner. He holds her favorite white lacquered tray with her rosebud tea pot, tea cup, and saucer trembling on it. He lowers the tray onto her lap and she clasps her hands together with joy. "You even remembered the slice of lemon! And honey!"

"And a scone!"

"A scone?" she asks, lifting the paper napkin from a small mound on the tray. It's an English Muffin, but she doesn't mind.

"A perfect scone!" she says, gazing up to him with tears in her eyes.

"Now, eat up, drink up—I'll get to work." He leans down, places his hand on the back of her neck, and kisses her cheek. "Thanks, sweets."

After Elinor finishes her "scone" and sips the last drop of her tea, she sets the tray down on the floor. Arthur practically levitates in his chair, full of energy. His faith in her worries her a bit, but she does take comfort basking in his gaze—it soothes the ache in her head. She begins. Click-click. Yarn twists in her lap, the strand in her needles shivers. Click-click-click-click. Arthur stares as an azure haze gathers around her—she doesn't notice.

He feels the change coming on. His kneecaps loosen and his quadriceps contract. He smells the lemon tea in her teacup. The scent makes the hairs on his arms stand on end. He's ready. He begins. Scratch-scratch-scratch-scratch.

Click-click-click-click.

Elinor hears a humming. Observes the clutter of sugar cubes, a lemon sliver, the curve of the steeping tea pot. She senses an inkling, a feeling, now, of frantic, roiling, microscopic behavior of charms and muons, of quarks. She gasps.

Teacups quake, spoons tink-tink saucers. The hum grows louder. It's not a hum, it's a deepening silence—a terrible in-between silence: the silence in anticipation of a cut tree hitting the ground. She watches as, from her lap, a brilliant blue ball of yarn tumbles, slides, swings in the air, sweeps backward in the gently curving arc of space, and skims her ankle. She sees Arthur in the room's white gleam, his gray head bent over paper, pencil pulsing.

He frantically scratches the pencil to his paper. The brain to the hand to the pencil to the paper to the theory. In silence, the shining blue bird delivers the elegant, unifying theory into his ear, into his eager, glimmering mind. He looks to her. He knows. He knows that every particle in his body, every speck of light that lets him read and every packet of gravity that pushes him into his chair, he knows is just a variant of energy of one fundamental entity. Vibrations. Strings.

She glimpses the snapping shut of turquoise particles around her eyes. They bound into her hair and fizzle into the air like lights turning off in a house at night. She leans forward, tumbles from her chair, to the soft carpet on the floor. She feels the give of the floor, feels her body push through it. She hears Arthur's voice, like a yelp from a child, his footsteps, running. A momentary lull. The click of a red button. The alarm. The incessant alarm. She feels the heat of his hand on her back while his other hand cradles her skull and, in it, her mind leaps forward, backward, through time.

Arachne

Elissa Malcohn

Arachne scuttled up her web to repose under a canopy of dead leaf. A fly hung motionless below her, one she had fought patiently, waiting and vigilant for signs of struggle. A large catch, it had jerked and twitched almost out of her web, tearing gluey strands. Her poison had been all but ineffective.

She had woven in and out like a boxer, wrapping more silk around the body and dodging its frenzied twists. A fly draws great strength from terror. It had taken her lifetimes, it seemed, to subdue this one. A silk cocoon preserved it now, rendering it anonymous.

Between panes of storm glass in a south-facing window, Arachne hung suspended, her legs curled gracefully up and inward. Her eight eyes turned in toward dreams. The leaf she had dragged for half a day and secured above her shielded her from strong sunlight as she slept, hiding her from marauding wasps. Heat eddied about, trapped in the window.

*

She awoke in a hammock in an herb garden. Arachne opened her two eyes and started. Her arms grabbed the sides of the hammock and it swung wildly, and she shifted her body weight to still it. Yesterday she had balanced herself on a thin strand as naturally as breathing. Today tall spires of grass had turned into a wispy carpet beneath her. The hammock around her was a white weave and her fingers tangled nervously between its knots.

She lifted her arms and looked at her hands, the clouds beyond them puffy as newly-shorn wool. *Human.* Her thick raven hair contrasted with the macramé around her shoulders. Her skin glowed with bronze radiance; high cheekbones rested above an aquiline nose. Full lips parted where once there were mandibles and chelicerae with fangs. Her voice had deepened but held no less awe, and indeed more, than when she'd been a maiden in Hypaepa before her first transformation. "Human..."

Arachne slid carefully from the hammock, gathered her white gown around her, and surveyed the garden. A cottage of brick and stone stood in the distance.

Was it a dream? Is this still a Lydian city; did Athene only knock me senseless with her weaving shuttle? What place is this? She studied the veins risen on the backs of her hands. *No mere dream. I've aged.*

She trotted to the cottage, ungainly at first. Then she grew accustomed to the feel of two rather than eight legs in motion, to arms swinging through air. Climbing three porch steps, she knocked on the door.

A silver-haired woman, tall and ramrod straight, answered. Arachne stood nervously at the door, bunching her robe in her fists. She looked up at a seamed face and eyes with crow's feet radiating like sunbursts, then took in the plain gray sweatshirt and cotton pants.

Ice-blue eyes gazed steadily into Arachne's dark brown. "Yes?"

"Excuse me—I was wondering—" what to say? I found myself in your hammock and I don't know how; I used to be a spider. "I'm—lost," Arachne stammered. "I was wondering if I could just sit down for a minute and get my bearings. A glass of water. If it's not too much trouble."

"If it was too much trouble I'd have booted you off my porch." The woman motioned her in. "Sit down on the couch inside; there's cider." Her lips curled into the slightest of smiles. "Unless of course you'd like something stronger."

Arachne followed her inside uncertainly, and followed her pointing hand to a den. Surrounded by walls of dark wood grain, she let her gaze wander. Her hand rested on the plastic casing of a television set and she squatted to run her fingertips over its blank screen. On a coffee table by the couch there was a pushbutton phone. Dream images. She knew what they were and how they worked. Now that she had seen them, it was as though the gods had filled her with knowledge awaiting only the proper cues for its release.

She looked to her left, toward the brick fireplace. On its mantle there sat a bronze helmet engraved with ram's heads, and a stuffed owl beside it. The helmet was Pallas Athene's, the owl her sacred creature. Arachne stared back at the kitchen and tried to still her trembling.

Her host emerged with a silver tray holding two pitchers and two chalices and set it down on the coffee table. She poured apple cider into one chalice and handed it to Arachne. As she poured the other her den filled with the smells of honey, juices, and spice fermented to a heady strength.

"I'd offer you ambrosia, but that's a god's drink," the woman said plainly. "Sit."

Arachne sat.

Athene sat crosslegged opposite Arachne on the couch and rested her elbows on her knees. "I've deprived you of a hard-won fly," she continued. "I hope this cider is an acceptable substitute. As for being lost, you are in twentieth-century Massachusetts, in the New World. Does that ring a bell?"

Arachne whispered, "It does now."

"Good."

"You brought me back."

"Does that surprise you?" Athene raised her eyebrows, and Arachne caught a glint in her eyes that reflected a younger goddess resplendent in armor. One who had perfected her youthful beauty as assiduously as she'd honed her skills. "I am in need of a weaver of your rare talents." The goddess lifted her ambrosia in tribute, with a wink, and sipped.

Arachne barked a laugh. "The last time I wove for you as a human, you turned me to spider."

"The last time you wove for me, you squirt, you escaped my wrath by hanging yourself and *then* I turned you to spider. There's a difference. A live spider is of more use to me than a dead girl." Athene gazed at the wall behind Arachne and its wood grain began to warp under her scrutiny. The lines in the wood straightened into tight, parallel strands.

Arachne followed Athene's gaze and swung around on the couch. A golden glow enveloped the wood. A bolt of light shot into the den from the kitchen window and shattered into prisms. Sunbeams breached every window and streamed in, accentuating angle and contrast. Shadows vanished; even the floor underneath the couch lay illuminated. Beams of individual wavelengths looped and spun against the wall, zigzagging with great speed. Two wefts merged into the wood grain and thickened from the bottom up. One formed a border of olives, the other of flowers and ivy.

From behind her, Arachne heard the goddess murmur, "I'm also quite good at restoration..."

Arachne grew dizzy watching twin oscillations of light, colors blending finely as royal Tyrian purple metamorphosed across the visible spectrum. To the left lay Athene's tapestry. Scenes unfolded, for a second time, of the goddess's contest with Poseidon for her city of Athens, and of Cecrops, its first king. Of mortal trials emblazoned in the weaving's four corners. To the right there spun Arachne's work. She remembered her crafting hands, fingers deftly weaving scenes of Zeus cavorting as bull, as shower of gold, as swan with mortal maidens. Scenes of Divine follies. She had been a girl of poor parentage and no family to speak of, who had dared to challenge the Goddess of Wisdom at weaving. They had set up opposing looms under the wide eyes of nymphs and Thracian women.

Clouds spun, drawn into threads arcing over the women's heads. Arachne looked up, gazing into lines fine as silk that sped into the warp on the wall. Deep in her belly she felt the stirring of phantom silk glands, the exhilarating pull of her issue through six spinnerets and out of her body, her leaps across chasms as she left a guy line behind her, eight legs poised and ready to grasp...

Without thinking she rose from the couch and plunged her fingers into the beams above her. Her legs twitched as she tried to reach them to weave; she lost her balance and fell back into the couch's plush upholstery.

The intense light vanished as quickly as it had come. The room changed once more, softening as contrasts faded. Shadows gained entrance into their accustomed places. The two tapestries, woolen and heavy, hung on the wall where they had formed, as fresh as the day they were first made.

Sprawled and panting, Arachne blinked. She flexed her fingers. Athene eyed her curiously. Even in sweats, the goddess's regal posture was commanding, a beacon of calm by Arachne's disarray.

"For a moment—" Arachne panted. "For a moment, I was back there."

"I know."

"It was—" *Exhilarating*. Arachne shook the thought from her head. *But no, it was—* She knotted her fingers together.

"You don't have to tell me," Athene replied softly. "That was a rather small test. You've passed it well."

Arachne straightened her limbs and drew her gown around her.

"You've worn the bodies of many spiders over the millennia. They've taught you."

Arachne glanced nervously about her, searching for a loom, a spinning wheel. Her fingers ached to weave. Her skill was not a matter of pride now, but an addiction. Spiders given insects without the need to catch them still spin their webs. They ignore the vibrations of trapped food. They build their webs, dismantle them, build them again. *It was—* Breathing hard, Arachne retraced her steps, tried to remember if there was a loom in the kitchen. Weave, or die. *It was—everything.*

"Have you a loom?"

"Nonsense," the goddess said sharply. "This is a modern house, Arachne. This is the twentieth century." Eyes of ice, her face taut and lined as a matriarch's.

*

Athene had led the disoriented woman to a guest room and put her to bed. Now she looked past the dormers of her bedroom and studied the heavens. *First there was Chaos...* Old god, her people's first Divinity. "We have a matter larger even than you," she whispered into the darkness. Arachne, that vain, stupid girl, had lasted from spider's life to spider's life and grown. Now returned, a woman with a spider's knowledge, Arachne possessed more than just a weaver's instinct. So Athene hoped.

She found herself wondering what the other Arachnes were like. Were some of them male? Were some of them still spiders, or spider-equivalents? Athene had no way of knowing. This Universe, this Chaos, was a puddle, and she was the goddess of a puddle. Eons ago Athene believed this Chaos was all there was beyond the Earth and Planets, just as the girl Arachne had believed in nothing beyond that exotic end of the world called Hyperborea. And *she* had been self-educated. There had been villagers in her small town of Hypaepa who believed there was nothing beyond Hypaepa...

As Adam and Eve were to humans, so were Erebus and Nyx to those people known as Olympians. Eve was just as much myth—and a human woman as well, placed in Africa approximately 200,000 years ago. What the human Arachne shared with twentieth-century peoples was a speck of mitochondrial DNA, passed on only in the egg and inherited by all humans from the same woman.

Pallas Athene also possessed a speck of DNA, one leading to the Olympians' common ancestor—a peer of Eve's perhaps. While Eve's people had a proclivity for reproduction, Nyx's descendants possessed other means, evolving under different rules. They did not aim for numbers but for legend. Interbreeding with humans added numbers to human legend and kept the Olympian population small.

While descendants common to both people possessed variation in their DNA strain, such variation had been labeled a result of the "mutation rate" in a modern people unaccustomed to recognizing their ancient gods as co-inheritors of the planet.

Mathematics and physics had bridged in the modern world to yield a theoretical knowledge of ten dimensions. Physicists, constrained by the dimensions around them, tried to reduce those ten to the four they knew of: length, width, height, and time. Blinding themselves. Even now Chaos raced outward, like pursuing like, to the meeting of its counterpoints at a common focus: Chaos multiplied.

Funny that I am not curious about the other Athenes. She smiled to herself. She had worn the masks of beggar, warrior, crone, maiden, shepherd boy. In and of herself, she had been a multiplicity of Athenes, able to take any form she desired. Able to become all possible players in appearance, she had not thought her own counterparts to be any different from herself.

Provincial fool—what if you'd sprung from Dionysus in another world, your chromosomes steeped in wine? What sort of drunken wisdom would you have imparted then? Perhaps in another universe the Goddess of Wisdom was a mere concept, a mathematical theory. Perhaps the sperm carried mitochondrial lineage, to make her bursting from Zeus a natural act rather than one requiring Rhea's genetic soup smuggled into a goblet of ambrosia. Or perhaps her counterpart had merely sprung full-blown from particles colliding in a cyclotron.

The Olympians, despite their godlike abilities, did not possess the required knowledge that would untangle the multiplicity of Chaos when the universes met. For that reason, Athene brought her challenger in weaving back into human form. Of all creatures, spiders that are orb weavers move to the beginning from the end. They begin their webs at the outside edge, knowing to use all their silk, no more and no less, and knowing their reserves before they start to spin. They are fully aware of the mass they have available to weave. The boundary of the web is woven first; then comes the movement inward to the center, the focus. A spider's first instinct is to create the perimeter while having advance knowledge of the finished product's shape and size. In weaving, the end and not the beginning is the *fait accompli.*

We are at the perimeter of a large space, Athene mused, *and we are expanding toward its center.* What's more, the discoverers of the ten dimensions had also included, in their Theory of Everything, the fundamental building blocks of matter and energy and named them:

Strings.

If Arachne was worthy, she, like her counterparts, would not only be the supreme weaver. She would become the supreme Creatrice. Squinting into the night, Pallas Athene stood calmly with the grace of Wisdom, while the silver hairs on the back of her ivory neck began to stand.

*

"Get up!"

Arachne's face lay partially obscured under the fringe of a blanket. Athene grabbed a corner with one hand and whipped her cover off the bed.

"Bitch!" Arachne spouted, fully awake now and curled in a fetal position to conserve her body warmth. Cold morning air hit her skin under a thin nightgown. Shivering, she rolled out of bed. "Damn it, don't you tap mortals on the shoulder?" She groped for her robe. "Or send the smells of breakfast into their nostrils, or send a waking dream of Divine import?"

"Give me that." Athene grabbed the robe from her. The crone was dressed in a turtleneck and slacks. "Dreams of Divine import; did you think I was here to serve *you*? I brought you back for a reason and *not* the other way around." Bluejeans and a cabled sweater appeared in a swirl of mist and settled gently on the bed. "Wear these."

Arachne dressed quickly and followed Athene into the kitchen. Pancakes flipped themselves on a griddle. As Arachne sat gingerly at the kitchen table, the Olympian pulled a plastic carton of apple cider and an unmarked quart container from an old Frigidaire in the corner. "Or would you prefer orange juice?" she asked, setting goblets on the table.

"Fine," Arachne said numbly.

The rose tint of the apple carton flashed a bright orange with a corresponding change in fluid and text. This simple play with matter and energy was still godlike to Arachne. She would learn to weave much finer stuff.

As Arachne helped herself to juice, Athene poured ambrosia into her own goblet. The ambrosia heated in the cup and sent tendrils of smoke upward as Athene returned the unmarked container to the refrigerator. Plates flew from the cupboard to the stove as pancakes leapt up in mid-flip and landed in a stack on each. The plates, like discuses, gyrated to the table.

"Did you bring me back to show me parlor tricks?" Arachne picked listlessly at her food. "Why am I here?"

"You are here to understand the parlor tricks. And you had better eat, because after you've finished I want you to chop firewood. We may get frost tonight." Athene spoke with her mouth full, gesturing with her fork. Arachne fixed her with an incredulous stare. "You'll find an ax in the shed out back."

When they were finished, Athene took their dishes and flung them, spinning, toward the sink. The dishes met briefly, edges passing through each other, transparent. Where they met, a shaft of light speared upward and dispersed, and the dishes vanished before colliding with the faucets. Arachne shook her head and turned away, heading toward the door.

"By the way," Athene shouted after her, "I *do* tap mortals on the shoulder when I wish to wake them. How many spiders have you lived? What makes you think you're a goddamn mortal?"

*

Arachne spluttered curses older than the New World as she strode to the shed where piles of wood awaited her. Pausing at the door's rusted hinges, she leaned against rough wood beams and tilted her head up to the sun.

She had been a silver argiope, silver-haired and spotted, her legs banded with black. How many lifetimes ago? She'd spun a new web nightly, her old one eaten, a spiral with zigzagged cross-strands. She had known self-sufficiency for centuries on centuries, taking the forms of myriad species of spider. Hatching with other spiderlings by the hundreds, she'd left a desiccated egg sac and never looked back. She had begun to build small webs. She increased her silk glands' capacity to produce more by draining them, and her webs became larger as she aged. Mandalas all, gluey strands shimmering in the light, a pinwheel of delicacy with unmatched resilience.

She remembered the pull, that core of her being liquefied and then drawn solid through her and out. She recalled her spider's belly yielding, yielding, swollen with warp and weft. One silk for radii and another for the frame. She had knotted her world around her time and time again.

A male had grown to mate with her. He had spun a small web on the outskirts of hers and then twitched hers, asking to approach in safety. Legs against legs he had entered her, washing his seed over her storehouse of eggs. Spilling his body into her as surely as she'd spilled her body of weave out into the world. And she, spilling again, pushed her egg sac from her. Drained, the energy to replenish herself lagged, taken by her eggs and the hundreds of spiderlings curled like homunculi waiting to hatch. They had taken her mate's energy as well, feeding from the body she had eaten, after she had opened her chelicerae and bitten off his head. She had grown weak, greeting the tang of death, hanging—for a moment—on a knot of rope, a girl dangling before an angry goddess. Eons ago the girl's head had shrunk as her belly bulged, her arms grown inward and fingers lengthened. Branded into primeval memory, this image faded into death, as surely as her spider's death faded the moment she pressed her new spiderling's body against the edge of her egg, and burst it....

Arachne caressed her stomach with her hand. The other hand, resting near the hinges of the door, felt a gentle prodding, a sensation moving lightly from finger to finger.

A small spider negotiated its way across the wrinkles of her skin, leaving strands between her splayed fingers. She removed her hand from the side of the shed and watched it move forward and backtrack, over and under with singularity of purpose. The nerves in her skin cells fired, tickled by the pattern of scrambling legs. She watched the spider as it webbed her hand, felt tears leave her eyes and course down her cheeks, dropping down toward dead leaves on the ground. Her chest began to heave as she cried openly, sobbing. *Communion.* The Universe was cupping her in its hand while inside, Athene taunted her. Was she loved or abandoned? Why was she here? Why was she *human*?

The spider dropped from her, spinning a guy line as it floated gently to the spot where her tears had passed into the soil. She watched it climb across blades of grass.

The web in her hand glistened. She wiggled her fingers, feeling the pull of silky tethers. She placed her fingers in her mouth, slipping her tongue under the sections of web and drawing it into her. *Keep your sour ambrosia.* She swallowed and stepped into the shed, ready to chop wood.

*

Stoking the hearth fire at twilight, Athene listened to the sound of hammering outside and smiled to herself. Slender logs and flat pieces suitable for a frame had lain on one of the woodpiles. Arachne had spent the day in Athene's garden, smoothing and sanding branches from which to make rod and heddle sticks, and chipping a flat piece to serve as the batten for her loom. Self-sufficiency, once learned, is not easily forgotten—even when one changes form.

Wool had miraculously appeared in the closets. Reclining in the den, Athene tuned her mind to a force unnamed by humans, spying with a channel of Sight from where she sat on Arachne's activities in the guest bedroom. The other woman had driven nails into the frame and was dressing her functional if primitive loom. All Arachne needed to do now was spin as a human and let her instincts meld.

There was a knock on her door. Athene projected her voice toward the transom and called, "Who is it?"

"Courier."

"Oh, it's you." Athene snuggled into the couch and produced a mirror, angling it toward the kitchen until she saw her door reflected. She called into the mirror, "I'm old and creaky, Hermes, and I've got a dandy fire roaring in here. I'm not getting up off this couch; come in through the mirror." As she reached into the reflection, her hand distorted into a tiny counterpart that grasped the doorknob and turned. She opened the door and Hermes's small image grinned at her.

"Punctual as a returning infection," she muttered. She lowered the mirror to the floor. Hermes stepped through, growing to full size as he rose.

"Take care you don't step on the glass; it's seven years bad luck."

He buoyed up, wingtip shoes beating feathers at his heels. "Better?" Landing on Athene's rug, he made a gallant bow, reached for the mirror and returned it. He kissed her on the cheeks. "It's toasty in here. You stoke a good fire, Gramma."

She snorted. "I see you're still wearing your youth like a proper delinquent."

Hermes threw back his head and laughed, then dropped onto the couch and crossed his ankles on top of the coffee table. He stretched luxuriantly and batted his long eyelashes at his half-sister. "Youthful androgyny is in vogue, dear. I don't look good in gray; *you*, however," he added quickly, "add—how you say?—elegance to the art of wrinkles. Your face is a manifold of beauty…a road map to the soul—"

"Who's here?" Arachne stood at the doorway. She stared openly at the curly-haired man, then at his caduceus that leaned against the side of the couch.

Hermes leapt off the couch and strode to meet her. "I don't believe we've been formally introduced. I'm the god of vagabonds, rogues, and thieves." His hand lay outstretched. "And travelers, science, and messages between the worlds, of which there have lately been a plague."

"I am honored, Hermes; I've only heard of you until now." Arachne shook his hand warmly, and felt herself falling into his emerald green eyes. "A plague of worlds? Or a plague of messages?"

"Yes." He kissed her hand. "The more worlds there are, the more messages there are. *I* have connections, of course."

"Can you tell me why I'm here?" Arachne asked.

"Isn't it obvious? You are here to weave."

"It's not so obvious to me; I had to make my own loom."

"Ah, yes." Hermes looked pensive. "And your own wool, too?"

Arachne glanced at Athene. Athene said, "I always stockpile wool in my closets. I hate to think of all those moths going hungry."

"You see—" Hermes lifted Arachne's chin with his hand, his lips hovering near her own. "I say I *have* connections. But *you* make them."

Arachne shrugged. "With strings, yes."

"Precisely!" Lips met lips and Arachne felt a stirring in her abdomen as Hermes took her into his arms.

She heard his voice in her mind: *What would you do if you had glands that could make all the silk in the universe?*

She giggled under his kiss. *I'd never stop.*

Good. Very good.

He slipped his tongue into her mouth and she felt her belly liquefy.

And if you could think strands into existence? How many strands would you make to fill a universe? Could you stretch yourself that far?

I wouldn't have to stretch. I'd know.

"Hermes, you let me know if you're gonna sweat her here on my rug so I can get a good vantage point and make myself comfy." Across from them, Athene shifted daintily on the couch.

Hermes slowly disengaged himself and planted a brotherly kiss on Arachne's forehead. "We'll continue this later. Gramma here likes to take snapshots and sell 'em."

Behind Arachne, the loom was a presence calling to the back of her head. She coughed into her hand and grinned at Hermes. "If you'll excuse me, I'm working—"

"By all means."

Athene looked after her with eyebrows raised. She said to Hermes, "Your son Eros is a bad influence." Lifting his caduceus, she playfully smacked her half-brother on the rump.

*

Hermes was sprawled on the couch, in bluejeans and cowboy boots. Wings fluttered in place of spurs. A *Spiderman* comic book obscured his flannel shirt as he read. He sprang up, leaving the comic on the coffee table, and bounded to Arachne's room. If his half-sister had planted a dream in that young beauty's head, he was going to see what it was.

He passed her loom and studied a tapestry of pinpoints. Alternately they spread thin and clustered together, meeting in tiny knots tied into the warp. Subtle shadings of indigo held fiery touches of red and gold and, beyond that, black.

Carefully he stepped around skeins of wool and approached the bed. Arachne, black hair fanned over her face, lay on her side with her left arm dangling beyond her pillow, pointing. Hermes knelt beside her and gazed beyond her forehead…

She skated the beams of a web that pulsed with golden light. Cross-strands dwindled to vanishing points.

Squatting, securely planted on a tightrope, Arachne's hand closed around the beam and disappeared in waves of electromagnetic radiation. Tingling, she felt its pull against her and inside; even as she rested a web streamed forth from her, unwinding somewhere else. Someone else had built the web that held her; now she was building another, one that she could not see.

She felt herself as a body of stars. Of galaxies and clusters of galaxies where they had spread thin and then clustered into knots. Of pinpoints. At the same time she was tiny, ten trillion times smaller than an atomic nucleus. They were one and the same. The soul knows all dimensions. A mile-high spire of grass becomes the wisp of a blade.

She followed a glowing trail. Strands ran parallel to hers. Energy fields crossed her path, curving toward the focus. She negotiated a maze from the outside in, all roads leading to cohesion. Or to chaos, where webs intertwined and energies multiplied, to the point where they would burst through their own fabric and open a new pathway, a new web...

She saw Athene superimposed at the breakfast table, flinging dishes at the sink. Saw the dishes meet and pass through each other, saw a shaft of light spear their common space. As though the light had been a portal admitting itself to another dimension, photons taking a course of least resistance. Moving forward once more. Arachne skimmed over an electric current of 100 quintillion amperes.

"Black holes have eight legs," Hermes's voice called in riddle to her back. "A spider eats its web before it builds anew."

She was running now, toward the increasing light. Toward a meeting of all the weavers she had ever been.

Sunlight streamed into her window and fell on her hair, striping her face beneath with gold. Her eyes opened and she sat, felt gentle hands on her shoulders and grabbed them. She pulled them around her and crossed them over her chest. "Hermes—"

"Yes, dearest."

She turned around and pulled him to her. She buried her head in his chest, unable to speak.

"They are called cosmic strings," he whispered, rounding the curve of her ear with his fingertip. "Every universe is a web. The webs are going to meet, and tangle, and be re-cast on another plane or everything will turn back to Chaos." He added, "Old grumpy god, that Chaos. He was passé eons ago."

"What do I do?"

"Weave," he said simply.

"Look at me!" Trembling, she pressed herself harder against him. Her head was upturned by his shoulder, eyes gazing at the space beyond the ceiling. "I was nothing but a spider! Children have killed me without a second thought."

"What you said last night—"

"You were *joking* with me last night!"

"—was true." He passed his hands through her nightgown and felt her ridged spine. He brought them around and cupped her abdomen. "Your silk is energy."

Arachne felt a jolt through her stomach and twitched. "Hermes, what are you doing to me?"

"Letting you know your silk, so you can use it."

As though her veins had been opened, she lay under the Olympian's touch and filled with a force like the swelling of a river, like a solar flare. Soon the heat and cold combined, her body pulling in powers that drenched her. She began to laugh, giddy and drunk as Hermes caressed her belly and waist. An aura of red light rose from between her legs, spread through her nightgown and traveled to her head in shades of orange, yellow, green, blue, and indigo. Violet crowned the top of her head.

On impulse she reached out and pointed upwards. A blue spark shot from her fingertip and thundered against the ceiling, splitting paint and dropping a shower of plaster dust on her forehead.

A plate careened in from the doorway and smashed against the bed-post. Arachne jumped as ceramic shards fell against her feet.

"Hermes, get out." Athene strode into the bedroom and jabbed her finger at Arachne. She jerked her finger up and Arachne rose, closer and closer to the ceiling, lifted to hang limply over the bed. The woman groped for a hold and grabbed air.

"You don't climb a web before you build it." Athene's voice was deadly. "You don't perform parlor tricks before you understand them." She dropped her hand to her side.

Arachne felt herself begin to fall. She splayed her fingers, pointing to the mattress, and felt a rush of warmth travel up and into them from her stomach. Her heart pounded wildly, aching to burst through her chest as she continued to float, held by a shell of energy that pushed her up as she sent it down to the bed. She glimpsed Hermes's open mouth, his eyes sick with worry, before she blacked out.

The Olympian caught her and slammed his palm against her heart. Athene made a sound of disgust. "Just like you to give her the tools that will kill her before she knows how to use them."

"She's not dead."

"And lucky for you!" Athene swung around the room and eased Arachne back onto the bed. Hermes raised his hand to catch his caduceus as it sailed in from the den. "She's been preparing for her destiny since Hypaepa; of *course* she wouldn't realize what it is! And *you*, so eager to channel to that next tier of existence you'd let her burn herself out."

Hermes caught Athene's stare in his own. Beneath her rage, he saw envy. "You, of course, have the requisite experience so as not to burn yourself out," he said softly, bending down to tend to Arachne. "Would that you could only have the weaving skills that bested you before—and will do so again."

A hundred dishes smashed loudly in the kitchen. Athene stood calmly by the bed, arms folded, gazing at the loom.

"Of course, you've ruined her tapestry once before," Hermes continued, weaving his caduceus over the prone body. "Do it this time, and that's the end of us and humans both."

"I know."

Athene turned and looked down at Arachne's pale cheeks, rendered colorless by her exertions. She, Athene, who had given fire to Prometheus—who had taught the Cyclops and Hephaestos to forge lightning for her father Zeus—she, who had guided Odysseus safely through his journeys and back home to Ithaca, would pale to a shadow of her doubtful existence in these modern times. All because of the destiny of a girl, a waif with nimble fingers.

Hermes looked up, into her sad eyes. "As the mortals say nowadays, you're a generalist, my dear." He rose and planted a kiss on Athene's cheek. "Arachne is a specialist. This ain't the old country, Gramma." He grinned.

Athene spun on her heel and strode to the kitchen. Hermes heard chimes as he worked over his patient, glanced with his Sight and saw the hundred dishes reassembled, mended as their pieces rang together, their cracks sealed shut.

*

"Look deeper."

Beside Athene, Arachne sat on her heels in the garden. In her left hand she held a keystone-shaped piece of jasper, maroon banded with brown.

"I see crystals," Arachne said.

Athene smiled. "Then you are looking farther than a conventional microscope. Look more deeply and eventually you will reach the subatomic level. If you change a spin, an orientation, at that level, you change the structure of matter. You change the perception of matter. You have already done that, but did not know how, and it almost killed you. If you see what you are doing, you can control your strength."

Gazing intently, Arachne listened. The words of the Olympian lodged inside her, bypassing her resistances. She looked more deeply, striving to reach into the atomic nuclei of the crystals she held encased in the stone in her hand.

They sat inside a domed field. Beyond the rose glow of their bubble lay stalks and branches gleaming with snow and frost. Athene laid her palm against the warm, translucent membrane: an extension of her own energies reinforced by the churning of the earth below her, where Hades rolled in crimson veins of metamorphic rock. Appropriate, Hades. Darkness, she thought, like Chaos. *You will not mock me, Hermes.*

Sometimes Arachne worked with crystals and pebbles, sometimes with water. More often, she worked with wool. When she wove, her loom expanded in size under her touch. She thought at first that she was mistaken, that it was an illusion: one of Athene's lessons disguised as parlor trick. But it was Arachne's energy that went into her tapestry and into the wool that spun between her fingers. Her knots grew smaller and more widely spaced as the weft between them grew. Whether she wove individual stars or individual atoms, she could not say. She knew that it did not matter.

The loom eclipsed her window, stretching the length of the wall. Her skeins remained ample; her colors changed as she wove them, absorbing and reflecting lights other than the rays that fell around her.

She slept snuggled in the crook of Hermes's arm, and dreamt of Heaven. He told her that Mount Olympus was a molehill and she laughed. Pressing herself against him she could feel the powers that coursed underneath his skin, could explore with her mind the knit of his bones. Outside each cell nucleus, she found the double-helix inherited from his mother Maia and the nonhuman women before her.

She passed her hand up across his thighs, and between his legs.

His hold on her tightened as her fingers moved. He felt her spin tiny webs of energy around him and tingled. She ordered him, "Lie still."

"With pleasure," he murmured.

He stole into her mind and saw himself pieced together, a composite puzzle. He saw his eyebrows rise in segments.

She drew him into her, legs and arms wrapped equally around his body, hands and feet moving in a busy rhythm. He was wrapped in tingling, as a resonance encircled his groin and then his waist, rising to flush at his neck. He hugged Arachne to him, feeling sticky as fermenting ambrosia. Moaning lightly, he began to thrust.

He started to move his hands to cup her buttocks against him, and was paralyzed.

Strings of energy continued to wrap around him. His eyes grew wide and he gazed again into her brain. He saw himself again in segments, the melding of eight ocular visions. He explored the insides of her body and found hunger. *Good Gods…*

He tried to open his mouth to speak, when Arachne's lips closed over his own and he felt his strength sucked out and into her. *Arachne!* He aimed his call like a lance into her mind. *Arachne, STOP!!*

The fabric of webbing between them blotted it out. He began to struggle. Arachne, entranced, continued her work. She followed his outbursts and subdued them, binding him.

His caduceus was leaning against the bedpost. Through a haze of pain, Hermes guided a thin beam of will in its direction, arced out of the reach of Arachne's fingers. His stomach was searing hot, his legs turned to molten lead as Arachne leaned over him round and plumped, ruddy-cheeked and unaware, enslaved to instincts from another life. The caduceus clattered against the bed, banging out an alarm softly, then louder. Hermes grimaced at the ringing in his ears as the clanging sent vibrations up through brass.

Pallas Athene remained conspicuous by her absence. Hermes focused on her helmet and owl on the fireplace mantle, moved his mind into the helmet and looked out through the eye holes burned into bronze. Once more he saw Arachne transformed into spider, under his half-sister's guiding influence.

With an internal shriek he shot his willpower into the caduceus. It whipped through the air like a saber, cut into the web that sucked his energies from him and sliced. Its serpents writhed, a double-helix wrapping around the web that wrapped around him. The serpents hissed, their own fangs widespread, pulling the golden threads through them and back into their bearer.

Arachne stopped her motions. Hermes pulled his hands from her back and flexed his fingers, working circulation back into them. *Your own brother, Athene! Or would you have come just as I was about to be scattered to the winds?*

Arachne stared at him, arms resting lightly against his back, and blinked in confusion. Hermes let himself fall back onto the pillows, his sweat pouring into the sheets. His chest continued to heave. "Arachne—" he gasped.

She touched him gingerly on the thigh. "I thought I was about to—"

"Eat me alive," he said calmly, exhausted. "You thought I was a fly."

"I *what*?" She spied his caduceus on the floor, gleaming with a riot of color that began to fade. Hermes closed his eyes as his breathing began to slow. Arachne lay down beside him, holding her face over his. "Hermes."

"Athene wanted to prove her point that a specialist is unsuited for the job," he whispered, eyes still closed. "She wanted to give me a good scare, so she changed your perceptions from the inside."

Arachne made a fist against the bed. "How?"

"By making you work with media other than your own. She can work her will into you, through those things that are less familiar to you." He opened his eyes, saw Arachne's worried frown and ran his thumb across her lips. "No fangs...that's good. Keep using the loom and don't stray from it. What you can learn through anything else, you'll learn through that, and better."

"Let me dream what happened to you," she said.

He chuckled. "Too nightmarish. You don't need that."

"Let me dream it so it doesn't happen again. I was acting like a spider, and that *is* what I am familiar with."

*

Arachne wove. Her loom changed daily, hourly. Minute alterations, stitch by stitch, transformed her tapestry into a place where futures lay empty of substance until given form. What the Fates had performed in legend for mortals—Clotho spinning the strings of human lives, Lachesis weaving their destinies, and Atropos terminating their threads—Arachne performed for matter, for gravitational anomalies and gauge fields. What was gravity but a web, the force that bound all things to enter it? Where did the Big Bang come from, if not an entrance from another place, a web consumed to be re-spun?

When the wall was filled with Arachne's weaving, Athene's house began to expand. The loom multiplied into manifolds, dimensions compacted. A seasoned traveler, Hermes guided Arachne through them as she learned new pathways and maps. Once she turned a corner and her loom was covered with spiders weaving into and out of the wool, arachnids that did what she did. An infinitude of spinnerets pulling liquid into solid, empty into bursting. Moving the universes inexorably toward cohesion.

Athene, dressed in full armor, watched in silence. Her helmet was pushed back on her forehead, the owl at her shoulder come to life and blinking a sleepy eye at a weave of patterns too small for human perception. The Olympian's hair was bunned up inside the helmet, but her skin was smooth and milky, all age lines removed. Hermes raised an eyebrow at her.

"Perhaps I only need to feel new again," she offered, fixing him with an inquiring look.

Arachne's tapestry had begun to disengage from its frame, its tassels snapping in a gathering wind. Waves of iridescence curled around her.

Hermes made to call out to her before she disappeared in a swirl of fabric and light, when Athene placed her hand on his arm and shook her head. "She has to get out there first," she said, "before we can follow. And then," Athene added, "she will be a part of those things that form *us*."

Arachne was no longer conscious of them, or of Athene's house. Her mind traveled where her fingers could not, feeding lines out from inexhaustible reserves. However long it took the universes to expand before they met at the portal, Arachne would spend it producing the silk to accommodate the new structure, fusing it with the webbing of her counterparts. Their work would mingle, and hatch from the closed loop of their present web. The new web would shoot sidelong, the next Big Bang in all its manifold dimensions, through warp of space and weft of time.

Arachne could see the center's approach. She continued her spinning, working from the outside-in. Her identities merged as she manipulated the particles of which everything was composed. She was the magma of Hades and the lightning of Zeus; she was Athene's wisdom sprung fullblown and Hermes's flight between dimensions. She was human and Olympian and spinner, Demeter and Hera, Ishtar as well as Aphrodite, Mehezh and Brigit and Agni and Au Set. She was a string, a theory, an equation. She was a particle zipping through a cyclotron, a cosmic ray blinking through the Earth in a flash.

Mortals always were the last to find out about their chosen, woven destinies. The woman from Hypaepa raced, now, toward hers, with slight of hand and singular purpose. Back to the beginning, when all was new—with all knots securely in place, and a universe held fast.

Same Old Story

Lynn Pattison

We must assume several more dimensions.
– Explanation of string theory, PBS

I

This morning I walked wobbly-legged to the 7-11 for my coffee, too sick to spruce up and drive to the brass & glass shine of Starbuck's for French Roast and Strenuous Smiles. *Pregnant*. Even though I crossed myself each time I slept with that heel. What luck, right when I'd laid my hands on flight coupons; had my checkbook in the black! Life's a cheap casino, and no jackpot bells ever clang at my table. Back on my bed I watch the cat stalk a finch in the yard plastered with wet leaves. When the bird escapes I cheer and slug the pillow. Gwen always said if she ever found herself in this fix she'd eat spider webs and lye, ground window panes, whisk it all into a thick soup, cloudy with flour and cream. But this will be alright—I'll learn to knit. Yellow blankets and sweaters, soft as feathers, tiny socks. Nothing scratchy.

II

The cat scratches and paws the pillows on the couch, I rub her round black belly and wonder if she could be pregnant. When I start breakfast she criss-crosses the kitchen, whisking her tail, until I let her out to chase wolf spiders. I got plastered last night, can hardly get the beans into the grinder, make the coffee—when I do it tastes like old socks. Yesterday morning, Gwen, the Pit Boss, came in all dolled up in a puce sweater and wobbly black stilettos. Brassy blonde with cheap glasses. She took over my table and sent me out in the hot sun to give out pull-tabs and coupons to the retirees spilling from busses with the *711 Casino* signs on their cracked windows. I try not to notice their cloudy eyes and ugly tie-up shoes. Later I saw why: that handsome, high roller was back. He always sits in the same place, leaves his body guard, Finch, slouched against the nearest door frame. He's had quite a streak, and she's itching to get her hands on some of those winner's tips he spreads around when he hits the jackpot. Wants to add to the stash under her feather bed, or tally up new numbers in her checkbook. It was nothing to me, I got a break from the smoke-tainted air, and time to plan which brand of soup I'd fix for supper. It'd take a goddam Healer's laying on of hands to cure this headache, my face is the color of flour paste. I'm calling in sick.

III

My morning coffee seems off—taint of brass. The plaster above the counter has been crumbling for months, did some drift into my cup? My left sock scratches, constant itch at my heel. The sun shines through the window above the wobbly-legged table in spite of spider webs that criss-cross the screen, and unwashed glass. Not a yard away from the window, whisk of finch becomes a black cat paw snagging him. Mercifully, the kill takes place out of frame. I've burned the eggs and splattered grease on my sweater. Nothing's going to go right today. I could use some of Gwen's luck. She's won three jackpots in a row at the casino, and her soup never comes out clouded or too salty. She never found herself in a fix without a plan, or missed a chance for free coupons at the Piggly-Wiggly. She says it's her lucky birth date: 7/11. Maybe I'll drive over, have her lay her hands on my checkbook, sift my flour and grind my French Roast. Ask her to put a couple of finch feathers under her pillow for a month, before I knit them into my socks.

IV

This morning Mom pulls spider webs from the plaster above the coffee-maker. My sister, Gwen, pads around the table in her socks, black on the bottom where she ran out across the leaf-plastered lawn, rolled newspaper in her hand, to swat the cat she's named *Jackpot* away from finches at the thistle bag. Jackpot whisks indoors and hides in the knitting basket amongst eleven balls of yarn, where we'll find him later, his paws caught in one. Mom grinds her cigarette under her heel, asks me, without a hint of guilt in her voice, to drive to the store for half & half and some French Roast, and I don't complain because Tim, who stands behind cloudy fake glass at the cash register, framed by the cigarette packs and the brass lotto dispenser, is the guy who makes me go all wobbly—gives me that itch I just gotta scratch. Mom would say if I'm not careful I'll end up just like she did: seven months pregnant, alone, staring out the grimy bus window. Thinking if she lied about her age they'd hire her over at the new casino. She pulls a twenty out of the back of her checkbook, and hands me a coupon. I notice the feathered pattern of cross-hatched lines along her lips, the little pillows of puffiness around her eyes. She wasn't as lucky as I am. Next Tuesday Tim and I are sneaking out of town. This place is a dead end, and we can't go wrong—we've got a plan.

V

The moon pulls down, fixing its light on the feather tree. Gwen is a black spider, criss-crossing the window frame with eleven silk threads, that the rain will plaster to glass before morning like a wet sock. I whisk spilled flour under the worn rug, hand knit by French grandmothers before the war; polish the brass statue of the pregnant cat-goddess; rub each of her seven heads for luck. I'm itching to lay my hands on one of those casino jackpots, just once, a decent bundle of scratch. I'd leave the fortune-telling business and this rusty house trailer with it's coffee-stain paint, and take off for someplace where the sun shines. Momma found herself a lucky rabbit paw, got free of this grind and left. A few slices of pot roast in the fridge, and nothing but expired coupons and some Campbell's soup on the shelf. I'm not spiteful, but I think it's wrong she took my cloud-colored pillow.

Like Marriage

Jarvis Slacks

I'm not a scientist, or a theorist. Or a physicist, or a mathematician. But I love my wife. The universe is like marriage. Countless people through the ages staring at the stars and writing complex equations and impossible theories to understand the most important thing that is the most important thing. When I touched my wife she moved her shoulder and didn't want me to touch her. I hated what happened to us. Late, sitting on the front steps of my home, the sky a blue and purple that I didn't think could exist, I asked whoever made this whatever to give me some sense, some understanding. I needed it and deserved it. It's not fair the world is like it is.

The Dome.

It is white. Well, not exactly white. It is more like pearl. Yeah. Pearl. It is pretty boring, really. Like my marriage. It is like a Dome. Maybe like a gazebo. But there aren't any windows or anything. And it is big. Huge. Like, ten stories by a block. And the opening, which people just call the gateway, is easily half a block wide. And my marriage wasn't always boring. And it glows. Even at night. That is how people first saw it. It was glowing. It was shining into people's windows at improper mornings. 2 am, people would wake up and have this light shining in their rooms. So they would get out of bed and go to the door and open it and there, in the middle of a wheat field in Tennessee or South Carolina or Georgia, was a big, pearl, Domed thing with a huge opening, shining. I use to wonder if people knew when their world wasn't the same. If they could recognize it. My wife asked me that once, if I recognized the moment it all went to shit and I said, what? And she said, never mind.

The first three appeared in those states. Tennessee, South Carolina, and Georgia. Each one was in a field of some kind. Birds flew around it and dogs barked at it. At first. Dogs barked at it until the sun came up and then they sort of paid it no attention. Animals ignored it really. Most people, they believe that, when something wrong comes to this world, animals are the first to notice and take heed. But animals sort of did their regular things. It was us humans that couldn't leave it alone. (Am I an animal? Animal-like? Base? Primal? Ape-ish?) Upon close inspection, people noticed that the huge Dome thing was seamless. It was like it was cut from something. And it glowed but it wasn't hot. And the Gateway, which was the size of at least half a block, sucked all the light away. That was a scientific term given by scientist, later. After more, deeper inspections. Inside the Gateway, there was this black, bluish darkness. The black was from the light being sucked into it. But the bluish part was from this mist that oozed out of the Gateway. And the blue mist tended to move towards people, and pull them. Kind of. You could resist it. But if you walked close, it helped you a bit. Like a breeze on your back as you walk up a hill. Like my wife, Nina, like her smile.

The first person to walk into the Gateway was Dylan Peppers. He was a farmer. The Dome thing was in his field and he and the sheriff and about fifty other people surrounded it at about eight in the morning. Dylan Peppers was going to harvest his tobacco that day, and the Dome type thing ruined it for him. He wasn't upset, really. He figured he could sell the rights of his story to television and sell the huge Dome type thing to Hollywood or NASA or New York. He thought the city of New York would want it. The sheriff told everyone to stay away from it until he could call State Authorities, which would call Federal Authorities. But, while the sheriff went to his patrol car for his cell phone, Dylan didn't see the harm. It was obviously supposed to be walked into and everyone dared him. They said, I bet you won't take a look in there. And Dylan said the fuck I will. Fuck them Aliens. The consensus was that it was a space ship.

Dylan got his shotgun and his flash light and before the sheriff could stop him, Dylan Peppers, 47, was the first person to walk into the Gateway of the huge, pearl, big Dome-like thing. He never came out. The sheriff, a deputy, and two other men with rifles followed him. They never came out, either.

After that, everyone in the world was watching this thing, which was soon dubbed the White Dome. It was called the White Temple and the Pearl Shrine at first, but no one felt good about giving such positive names to such a thing that swallowed people up, unforgivingly. That's like marriage, too. What's not like marriage? Federal Authorities cordoned the Domes in Tennessee, South Carolina, and Georgia off to the public, but the media was all over it. The idea that Aliens were involved, you know, from outerspace, could not be shaken. But the Federal Authorities did what no one thought they would do. They gave all their information up as they learned it. Everyone thought they would lie or hide it or try to destroy it, but they didn't. Well, they did try to destroy them, but that was later. Anyway, instead of lying or hiding, they told the truth. They ran test after test and told the world everything they knew. The spokesman for the White House spoke more than daily.

"It isn't radioactive. We don't know how the light is generated. We can't tell you, at this time, where the five men are. We aren't sending anyone in there until we have more facts. No, we can't move it. We have tried. It is, seemingly, impossibly heavy. Well, yes, that is true. It should be sinking into the ground. But its not. The mist is some sort of chemical compound. We have tested it in the labs. Nothing conclusive. The Gateway of the Domes seems to suck all light into it, like a black hole would. But, in order to cause that effect, it would have to have an immense gravitational pull, which it doesn't. We swear to the people of the world that these things are not the property or cause of the United States government. Personally, yes, I would like to walk inside."

Everyone did. Worldwide, over 1/3 of people polled said they wanted to walk inside of the Domes, whether it hurt them or not. The Government took action, using the National Guard to block the Domes, making it impossible for anyone to enter it. At one time, in the field in South Carolina, the massing crowd was numbered to be about 100,000.

After three days, they sent a bomb robot, the kind of robots they use to defuse bombs, inside the gate. They tied a 20 test ton cable to it, and attached the cable to a large transport vehicle, tethered into the ground. The robot rolled inside. After five seconds, the cable snapped and whirled back out, the end cut. Technicians said the cut was so precise that it was almost atomic, cut between the protons and electrons. That's it. That was all the government needed. No one was going inside or near the Domes until more investigation was performed. The cordon around the Domes was widened to twenty miles in diameter. Whole counties were evacuated. Better safe than sorry. The next day, one thousand more Domes were reported, worldwide.

At the time, I worked for SINT, the Scientific Institute for New Technologies. SINT was government subsidized, but not owned. Meaning, they gave us money and asked us politely to do things. The cool thing about SINT was that they could skirt media attention and public outcries by saying that they were a private corporation. And the cool thing about my job was that I didn't need to know dick about science, institutes, or technology. I just had to talk to people. And travel. I wish hindsight were foresight. I can see how my marriage crumbled like a house made from saltine crackers.

"What did you say your name was?" a lady asked in Seattle, Washington, a days drive from where Dome #67 appeared.

"My name is Jon Filmore. I'm with SINT. Here is my card."

"OK."

"We are part of a government collaboration to study and understand the White Domes."

"Are you a scientist?"

"No. I'm an investigator."

"Are we going to die from these things?"

"I don't think so. I can't say yes or no about that, ma'am. But, I mean, I doubt it. They don't look like bombs."

"Are you married?"

"Yes, ma'am."

"That's a shame. I would love for you to meet my daughter."

Whenever I interviewed someone, the subject of my marriage came up. Maybe it was how big my wedding ring was. Gold and thick and almost as wide as a nickel. It could have been how worn out I looked, around the eyes. Married men sometimes look like semi-beaten children.

"What do you think you'll do with your Dome?" I asked an old man in Detroit, a White Dome in his front yard, his grandchildren shooting it with water guns.

"It's all bullshit, like the lunar fucking landing," he said.

"That doesn't really answer my question, sir," I said.

"You're married, aren't you?" he asked.

"Yes, sir."

"I see that same look in your eye I saw in the mirror, back in '42," he said.

"What look is that?"

"Supreme servitude," he said.

"I love my wife," I said.

"You can love them," he said. "But it's a whole different cow patty to like them."

I thought about that the whole trip back home, the whole plane ride, the whole taxi ride to my house. I tried to remember that moment, that moment I first saw her. That moment of amazement and wonder. But it was blurry, barely there. I could hardly grab hold to it as tight as I wanted. I tried to bring this up to her, when I saw her. My precious wife.

"Try not to talk to me," she said when I opened the door, arms opened wide. "I either have a headache or a stomach ache. I can't make up my mind which one."

Nina Filmore.

My wife liked for me to bring work home with me. And I did. Charts and files and statistics.

"What are those?" Nina asked. The oven was always cold, the kitchen table always bare.

"Charts, files, and statistics," I said.

"More Domes?" she asked, brewing coffee.

"It is nine at night," I said.

"My own personal clock," she said.

"Do you have any idea what caffeine does to your system?"

"I'm tired of systems," she said. She had this way about her, this way of saying things that would make my insides revolt and shiver. There were times when I wondered why she said anything to me, why she came home, why she kept certain parts of the house clean. Why she showered. Why she wore her wedding ring. Why she breathed. A husband should never have so many unanswered questions towards his wife.

"Charts, files, and statistics," she mocked.

"Charts, files, and statistics," I said.

"Attempt to impress me," she said.

"This chart," I showed her a multi-colored spreadsheet with horizontal numbers. "Shows the Domes in relation to the population that lives around them."

"There's a Dome in New Delhi?" she asked.

"Yeah,"

"Hey," she said, pointing at the file. "There's one in Moscow."

"There's one in every major city," I said.

"What's in the folder?"

"The United States' attempt to nuke the one in Arizona."

"I heard about that. How did it go?"

It went horrible, I told her. A megaton something or other plowed into the Dome in the middle of the desert and it went Boom, big Boom, mega-Boom. And when the mushroom cloud blew downwind and caused radiation poisoning to three cities in three states, the Dome was still there, still bellowing blue mist, completely unchanged by the mightiest force man could ever cook up.

"What's it like in there, you think?" she asked at that moment when you are in bed, about to fall asleep.

"No one knows."

"I said, you think."

"What do I think?"

"What do you think?"

I pretended to fall asleep before I could tell her. I pretend to need to be up early for work. I pretend I have to work late. There's a whole group of people that think the Domes are pretend, too, that they are all bullshit and make believe. But, every time I interview a detractor, a disbeliever, they would lower their heads, walk me to their back yards and show me a little mini, personal Dome, the size of an outhouse but just as bright, just as pearl. They were beginning to come in different sizes. At almost every preschool, elementary school, junior high, and high school and college, there were house size or car size Domes in the playground or common area or parking lot. Churches had them pop up in their graveyards.

Hospitals had them on their rooftops. One huge Dome, with an opening large enough for cruise ships to go into, floated in the Atlantic. There was a rumor that a Dome was floating in the sky and someone at the lab said high-powered telescopes caught a glimpse of one on the moon. And people were walking into them.

At home I would walk around the house before I went to work, look in the extra bedroom we had. We put junk in there. Old boxes and unused exercise equipment and folders and old clothes. It was supposed to be for a baby. It wasn't supposed to be a spare room. At work, we would sit around a huge table, with all those papers and computers and pins and data and we would spend days talking it out, trying to make sense of it all. But if we started at eight in the morning, we would stop at six or seven, close to sunset. We would all gather on the roof, and, throughout the world, everyone would watch the sun go down and try to count the Domes, the glowing white monsters. And each evening, without failure, there would be more Domes; the world would be brighter with every passing day.

I took a left instead of a right and ended up at the public library. The internet can never replace the book. The attainment of knowledge from wood, the paper, that old dusty stuff that smells of the forest, the earth. One book after another I pulled from the cases, placed gently on the table, and searched through. I needed knowledge that I didn't have. Worm Holes. The bending of space to create a portal, a movement. Was that what they were? Little portals to send us to a better place, where the sun isn't as hot, where we don't treat each other as harshly? I touched a Dome, once, even through the government had given strict instructions restricting the touching. The white walls vibrated. A harmonics. String Theory. Instead of the silly little balls that they teach us about in grade school, instead of atoms and particles and orbs revolving around orbs, there are strings, these tiny strings vibrating to give us the illusion of particles and orbs. The illusion that there is something there that isn't. Like my marriage. An illusion of a man and a women bouncing around the home they share, the illusion that things still exist as they did, even though everyone knows otherwise. Vibrations simulating particles. Making things seem what they aren't. I flipped page after page, definition after definition, becoming frantic at times. Theory of Everything. An explanation why everything works, how it is tied together, how it is held. Those scientist have it wrong, I thought. They should study why everything, always, tends to fall apart.

"We should be afraid," Beth, this girl I worked with, said to me, on the roof during sunset, watching the glowing world.

"But we aren't," I said. "What does that mean?"

"Don't most people feel this absolute calm before they die?" she asked me. I shrugged my shoulders. She started to cry these huge tears. I never saw her again.

I slept at the office four out of the seven nights of a week. I liked it there. I liked the hum of the A/C, the noise the elevators made at three in the morning when the cleaning crews came. This, of course, upset my wife.

"This marriage is ending," Nina, my wife, said over a dinner of frozen fish sticks and kettle baked chips. "For no other reason than that you want it to end."

"What are you talking about?" I asked. We used to watch television while we ate until the only thing the TV focused on were the Domes, the huge gateways, and the masses of people waiting in line to go inside. I tried to think of her how I used to think of her. I tried to remember the first time I ever thought of her, ever wanted to sleep with her. I used to like sleeping with her. It was something I enjoyed. Now sleeping with her made me want to tear my skin off. When did that happen?

"I'm talking about the end of our married life," she said, staying focused. "I'm talking about the stopping of us as a couple."

"I'm gone half the time confirming reports. How can our marriage be ending without me knowing?"

"Exactly!" she screamed. "How can you let that happen?"

"Why do I have to be blamed?"

"Who do you want to blame? The Domes? Those big escape pods from heaven?"

"Who said they are escape pods from heaven?"

"You are so dense you sink in swimming pools," Nina said, throwing her dinner up in the air and storming out the house, leaving the front door wide open. A maelstrom of processed fish, refried potatoes, and emotions spilled everywhere.

The world was leaving. In droves. They were leaving in packs. They were basically skipping into the unknown blue abyss. Like every single human being secretly wanted to commit suicide but didn't want to say, then the Domes and a reason for the leaving. An excuse to be gone. A quiet way out. There were reports in the poorest parts of Africa of entire villages, once a Dome appeared close to them, walking into them. Hundreds at a time. In Jerusalem, a Dome appeared there and a huge war broke out over the rights until the next day, when about three hundred more Domes popped up in the region and then every Muslim and Jew just ran inside them, hoping to beat the other to the everlasting whatever. Domes showed up in prisons, in ghettos, in hospices. Bored rich people walked into them. The homeless walked into them. Jaded black people, jaded white people, jaded anybody walked into them. Artists, musicians, writers, teachers. A school in upstate New York hosted a "leaving day" and guided the entire enrollment of a school into them. The murder rate sky-rocketed for a while because people would kill other people and just run into a Dome to escape. The president of the United States gave a speech to the people not to give up hope on this world, that it was not completely corrupt, that it was savable and then, in mid sentence, a Dome appeared next to him the size of a Port-a-Potty and the president of the United States yelled, Thank you, Jesus, and jumped into it head first. There were too many Domes now to stop people from walking in and whenever someone even wanted their own, private Dome one would appear within minutes. At night, there was no night, only the amazing white light of the pearl Domes, growing brighter with each prayer to whomever, with each final push to give up and leave this place.

Researching the Domes wasn't really interesting anymore. They may have caused the exodus of half the population, but they didn't really cure any diseases. And they weren't doing anything bad. They were just there. So, of course, SINT let me go. But gave me a pension big enough so that I would never worry about money again. I asked why and they said why not. Before I left my office, a census tech estimated that, of the 6 billion people that were here before the Domes, 3 billion where left. Life was less. The streets were bare, parking a breeze. Once the chaos of a missing workforce eased, the economy balanced out. And then life sort of became ordinary again. The Domes made no more news and very little people walked into them. One theorist believed that anyone who wanted to leave already left and now the world was full of people who liked it here. I heard he walked into a Dome the next day.

I sat on the front porch of my house most times, sipping green tea, listening to all the music I never had a chance to listen to, reading all the books I never read. I sort of liked the Domes. They were new things on an old world. One morning, I went to cut the grass and I saw a small Dome in my backyard.

"Did you wish for that?" I asked.

"For what?" Nina asked back, drinking tea in the backyard, looking at it.

"You know what."

"Maybe."

"Do you hate me?"

"Maybe."

"Are you going to walk into it?"

"Maybe."

"Do you want to leave?"

"Maybe."

I picked up, with a huge strain, a lawn chair and threw it at the Dome. The lawn chair went right into the gateway, blue mist bellowing out.

"Why don't you go get it?" she asked, laughing, laughing so hard she spilt her tea and fell to the ground.

What happened to us?

I tried to remember her, when I first saw her. The furthest my brain would go was back to when we bought the house, I got the job, we tried to make a life together. Was there a life before her hateful remarks? Her ill temper. I got in my car and left. As I drove off, I saw Nina walk into the middle of the street, watching me leave.

My parents lived towards the mountains. About a five hour drive. It was deep into the night when I got to their house, even though it wasn't really dark anywhere anymore. My mother was out of the house as I parked, wiping her hands on an apron, moving a stray hair out of her face. She was rounder than before. She had make-up on.

"I was just about to call you," she said, kissing me when I came up to her. She looked different. Something was in her eyes.

"What is it?" I asked.

"Where is Nina?"

"Home," I said.

"You two alright?"

"No, we're not alright."

"My baby," my mother said, touching my face. "Come into the house. I've made a huge meal. We've got company over, too."

"For what?" But I knew, before she even said it.

"A Leaving Party," my mother said.

My father was in the middle of the crowd. There were at least sixty people in the living room. Eating, drinking, laughing. They were all dressed in fancy suits, their hair done up, fine jewels on their necks and on their wrists. Parties like these were common. Huge gatherings where people got together and gave eulogies of the living. And then everyone would march into a Dome. I walked into the kitchen, pulled the curtains back. There it was. A big Dome, too. About the size of a large shed. I turned and looked at my mother.

"Don't try and talk me out of it," she said.

"I won't."

"They're saying that they're gifts from God," my mother said. "Life boats before the end times. Do you believe that they're life boats before the end times?"

I looked at my mother, put my hand on her shoulder. She wasn't well. Something in her head had snapped. People like that, your body knows it. You can tell when something in a person's mind has just stopped working.

"Do you believe that?" she asked.

"Yes," I said. "Yes I do."

"Will you come with us?"

"No," I said. My mother then made a small sound, like air bursting out from someplace special. She grabbed a tray of finger-snacks, and I heard my father say a joke. I could tell it was a joke by the cadence, the way his voice was working. My mother went to the living room and I went back outside, got in my car and drove back home. Five hours straight stopping only once for gas.

When I got back home, it was late, but not dawn. Nina was on the front porch, holding a cup of coffee, reading a thick book. She didn't move her head when I stood next to her.

"I love how you use avoidance to handle situations," she said without looking up. "Remember that time in the city when you thought we were being followed? You made us sit in a diner for two hours."

"I'm cautious," I said.

"You're boring," she said. "You're a coward."

"My parents are walking into one," I said. Nina put the book down. Her face melted a bit. For the first time in years. An almost tenderness. She tried to stiffen but then softened again.

"I'm sorry," she said.

"Why do you hate me?" I asked. Sat down on the steps.

"I don't know," she said. "Sort of sick of you, I think. You know how we clean the house? Fix dinner? Talk?"

"Yes," I said.

"I'm sort of sick of all that."

I got up. I took a long shower with the water as hot as I could take it. I put on my favorite pair of jeans, an old college sweatshirt. I made a hamburger. I ate it slow. I swiped the oil from the plate with my fingers and licked it. When I was done, I found Nina still on the porch, the book down and her face wet.

"I'm going," I said.

"In there?" she asked, pointing to the back.

"Yes."

"Are you sure?"

"One of us has to."

"We can't just get divorced, right?" she asked. "You can't just move away, right?"

"Seems sort of half-assed, now," I said. Walked to the backyard.

The Dome was white. It was simple and almost elegant. It was almost the most amazing thing I'd ever seen. But not quite. When I first saw Nina. Standing in the store holding the day's newspaper in one hand and a pack of cigarettes in the other. That coy smile she gave me.

"Buy my shit?" she had asked me. I looked away from the Dome and turned around. She was there, looking at me, her arms folded. I turned back around and jumped in.

"Why should I?" I asked her back then, the first time I met her.

"Because I asked you to," she said, that smile, those eyes, that play with her shoulders.

"Ok," I said. She put the cigarettes on the counter, then the paper. I threw down two bucks and told the cashier to keep the change. We walked outside and it was freezing. So cold. I put the hood of my coat over my head and she scrabbled to get the scarf around her neck. She put a knitted hat over her short hair. She took out a cigarette and handed it to me.

"You smoke?" she asked.

"No," I said.

"You're gonna have to if you want to date me," she said. The laugh she boomed out. She got spit on the face. She quickly wiped it off.

"You don't want to date me?" she asked.

"It's a little presumptuous," I said.

"Presumptuous," she said back. "Ah. That sums me up completely."

She lit her cigarette, then handed me the lit one. I took it and she lit another. I inhaled. The smoke was thick and warm. Moist. It tasted awful, but once I let the smoke go into the lungs willfully, it was love. The whole body was warm.

"Nice, isn't it," she said.

"Yeah," I said.

"Some bad things aren't all bad," she said, then extended her hand in a quick jerk. "I'm Nina."

And hanging there, swimming in the blue mist of the Dome, that's what you got. Those few minutes of absolute joy that you received in your life. M-Theory. The Theory of Everything. Closed Strings, looping together back where it all came from. That's what you got. Then a poof and a suction and a movement like falling from a great height, only you're falling sideways. And the blue mist gets thinner and thinner and you become less and less of what you were before and then you are something utterly unconnected…

Too Many Yesterdays, Not Enough Tomorrows

N. K. Jemisin

The alarm clock buzzed at seven, right after reality rolled over. Helen tapped the snooze button for ten more minutes. When the alarm went off again, she believed for a moment that a man was in the room creeping toward her. She sat up ready to lash out with nails and fists and feet, then memory returned and she chuckled to herself. A dream. Habit. Too bad.

#

BLOGSTER login: Welcome, TwenWen!
[Thursday, ??? feels like 10 p.m.]

Hel, you had the rapist dream too? Thought I was the only sicko! Y'know, back in college psych they said those kinds of dreams are a representation of your subconscious yearning to be rescued from your out-of-control situation. (That, or you want a penis. ^_-) Usually I try to keep mine going awhile, see if he actually manages to score. Never does. Figures; even my Freudian fantasy rapists are pissant schmucks.

In browsing news, surprise! There's yet another spec-thread running among the *BumBloggity brats*. "The government did it" version 2,563,741. Wish they'd get back to aliens or God; those are more fun.

BTW, gang, meet *SapphoJuice (his blog)*. He's in a snowy reality. Has a studio, poor guy.

Hey, anybody heard from MadHadder lately?

#

Life, post-prolif: she climbed up from the futon and shuffled across the room, her feet chuffing along the tatami-matted floor. When she reached the kitchen she took care to yank the fridge door open so that the glass bottles would rattle and clink. Noise made the apartment seem less empty. Then she slapped onto the counter the items that would comprise her breakfast: a cup of yogurt and a cellophaned packet of grilled fish. She rummaged awhile for the stay-fresh drink box of chai tea concentrate; she knew where it was, but rummaging helped to kill time. The milk was as fusty as ever. Irrationally she always retained the vague hope that if she got up soon enough after the rollover, it would taste fresher. Mixing it with the chai covered the not-quite-sour taste, so she microwaved it for three minutes and then used that to wash the fish down.

Chewing, she paused and grinned to herself as she felt a bone prick the inside of her cheek. She'd eaten the packet of fish seven times lately without finding it. The bone was always there, but tiny and easy to miss. Finding it made her feel lucky.

It was going to be another beautiful day in infinity.

#

BLOGSTER login: Welcome, SapphoJuice!
[Cinco de myass, the year 2 bajillion and 2]

SPINNYSPINNYSPINNY

Hi, all. Thanks for the warm greetings. My daily routine includes two hours of spinning around in my desk chair. My mom never used to let me do it before, so...whee!

Yes, Marguille, your guess as to the origin of my username is correct; I am indeed a squealing Herbert fanboy (sorry, Conty, not a lesbian =P). Only got Children Of in my studio, though. Sucks donkey balls. Big hairy fat ones.

Ah, c'mon, Twen, specthreading is fun and oh so good for you. Granted, it's pretty much a complete waste to wonder how and why the quantum proliferation occurred because we can't do dick to fix it...And granted, the *BumBloggers* do seem to have the same arguments over and over (and over and over) again...but hey, there's comfort in the routine. Right? Right? ::listens to crickets::

Hel: wow, Japan? You must have been quite the adventurer, before.

#

Jogging; she loved it. The rhythmic pounding of the hardpack under her sneakers. The mantra of her breathing. She would never have taken up jogging if there'd still been people around to watch her, maybe point and laugh at the jiggly big-boned sista trying to be FloJo. Before the prolif she'd only just begun to shed her self-consciousness around the Japanese.

They rarely stared when she could see them, and her students had gotten used to her by then, but on the street she'd always felt the pressure of the neighbors' gazes against her back, skittering away from her peripheral vision when she turned. The days of Sambo dolls at the corner store were mostly over, but not a lot of Japanese had seen black people anywhere except on television. *My parents must've felt the same during grad school in Des Moines*, she'd always told herself to put things in perspective. It hadn't helped much.

Now, free from the pressure of those gazes, she could run. She was fit and strong and free.

Around her the barren, cracked desert stretched unbroken for as far as the eye could see.

#

BLOGSTER login: Welcome, KT!
[Saturdayish, The House That Time Forgot]

Fighting the lonelies. Everybody still out there? Conty? Guille? Hel? Twen? (Hi, Sapp.) I haven't heard from Mad-Hadder either. What if the silence got him?

Don't want to think about that. Topic change. Did you know Mr. Hissyfit keeps going through the rollovers, too? I guess cats *do* think.

Sappjuice, it sounds like you're living in Fimbulwinter (sp?). I've got grassy plain. It's boring, but at least I know it can't kill me. You have my e-sympathies.

#

She liked best the fact that the day started over after about ten hours. Incomplete reality, incomplete time. She'd stayed awake to watch the rollover numerous times, but for a phenomenon that should've been a string theorist's wet dream, it was singularly unimpressive. Like watching a security camera video loop: dull scene, flicker, resume dull scene. Though once the flicker passed there was grilled fish and stale milk in her fridge again, and her alarm clock buzzed to declare that 7:00 a.m. had returned. Only her mind remained the same.

She usually went to bed a few hours after the second alarm. That gave her time to print out the latest novella making the rounds in cyberspace, read it in the bath, and maybe work on her own would-be masterpieces. It didn't bother her that the poems she wrote erased themselves every rollover. If she wanted to keep them, she posted them online where the mingling of so many minds kept time linear. But doing that exposed the fragile words to the scrutiny of others, and sometimes it was better to just let them vanish.

She decided to post the latest one to share with her friends. The new boy wasn't a friend, not yet, but maybe he had friend-potential.

#

BLOGSTER login: Welcome, Marguille!
[Sunday, 5 Marguille'sMonth, 2 years A.P., 2 a.m.]

I agree with Twen; specthredding is evil. But I can't help it; been reading the *Bumwankers stuff* (I know, I know). my vote has always been for the government theory. $87 bil. for an "emergency fund"? Shyeah. Probly only took half that to build some knd of new super-weapon, or hotwire a particle acelerator. "I know! Let's shoot some protons at the terrorists! Yeah! Oops, we bro,ke the universe!"

But seriously...I keep thinking that somewhere out there, normal reality still exists. no, scratch that—*I know* it exists, because it's possible. Fun with quantum theory! 'Course, that means oblivion exists too. (This is what we get for letting that guy Shroedinger experiment on his cat. Should've sicked PETA on him.)

SappJuice, don't feel bad about your studio. Hel's Japanese apartment's probably half the size of yours. (What do you call half a studio? A closet? ::ducks rotten tomatoes from Japan::) Anyway, it's not like the rest of us are so much better off. What difference does a few square feet make when they're the same square feet every damn day?

#

She got the email just before she would've gone to bed. The ding from her computer surprised her. Weblogs worked, as did other forms of public communication. Direct, private contact was impossible. Individual-to-individual relays—instant messaging, email—worked, but were always iffy. Most people just didn't bother to try; too disappointing. And then there were the rumors.

But she read the email anyway.

"To: Hel
From: SapphoJuice
Subject: Hi

Helen (seems so weird to say your full name),

Hope you get this. I read the poem you posted in your blog. I just wanted to say…it wasn't beautiful, but it did move me. Made me remember the way things used to be, and made me realize I don't really mind that the old world is gone. I got put in a garbage can by football players *every day* during my freshman year. My mom always used to tell me I'd never amount to anything. How could I miss that? Anyway.

I guess the only thing that bothers me now is the silence. And sometimes I don't even mind that, but sometimes the snow just gets to me. Why the hell couldn't my pocket universe have formed around an *interesting* environment? I could dig an endless beach, maybe an endless forest. No, I get snow. It's so quiet. It never stops falling. I can't go out far without losing the apartment in the haze. Sometimes I want to just keep walking into the white, who cares? Then I read your poem.

Sappy (yeah, I know)"

She sat at her computer savoring the newness of the moment.
#

BLOGSTER login: Welcome, KT!
[Ohwhocares? Someday, somewhen]

Mr. Hissyfit got out. I tried to catch him but he just ran straight away into the grass. I keep going out to call for him, but he must be too far away to hear me.

Stupid cat. Stupid goddamn cat. I can't stop crying.

#

She emailed SapphoJuice back and told him that she had only feared the silence once. That had been right after the prolif, when she'd still been adjusting. She'd started running and hadn't stopped; just put her head down and cranked her arms like pistons and hauled ass as fast as her legs would take her, as far as her lungs could fuel. When she'd looked around the apartment was gone, swallowed into the cracked-earth landscape. Instant panic. The apartment was only a fragment of reality, but it was *her* fragment of reality, her only connection to the other incomplete miniverses that now made up existence. Even before the prolif she had been happiest there.

She could admit that, now, to him. But back on the day she'd run too far she'd been in a panic, her grip on sanity slipping by cogs. It had taken the threat of true isolation, of wandering lost through endless wastelands until thirst or exposure killed her, to make her see the apartment as haven and not prison. So half-blinded by tears she had run back, thanking God that her shoes were cheap. One of them had an uneven sole which scuffed a little crescent-shaped mark into the dusty soil. The moon had led her home.

#

BLOGSTER login: Welcome Conty!
RED ALERT
[Day 975 (yeah right I actually keep count in my head)]

KT no more kidding. Fight it. Don't think about the damn cat. Go out and run—you can go pretty far from your house in the grass, can't you? Eat something. Hell, eat everything; it's not like it won't come back at rollover.

Talk to us.

#

The emails she sent didn't always go through. More than once she had to send them again when they bounced or, more often, simply never got a response. She saw the bounce histories in his attachments and knew that he'd had to send his multiple times, too. Just another day post-prolif.

She did not tell the others about the private correspondence, and neither did he. She knew what her friends would have said. It became something special, secret, a little titillating. As the days passed her dreams changed. Now the man creeping about her room had a face and a much less sinister demeanor. Now he looked like a skinny, geeky teenager, whose shy smile was for her alone.

#

BLOGSTER login: Welcome, Marguille!
[Jan. 37 errordate errortime 12:5g0k p.m.]

SILENCE.

You guys want to chat? I need some facetime. I think KT's gone.

#

Over the exchanges she shared her life story with him. Growing up less than middle-class, trying to act less than upper-class. The teasing in elementary school because she "talked proper" and couldn't dance. Her first boyfriend, a white boy—she'd been too guilt-ridden to bring him home to meet her parents, and they'd broken up because of her shame. Her next boyfriend, the one she'd almost married until she found out he was cheating on her. Graduating college and feeling the isolation grow in her life. Few friends, none of them local. No lovers. She'd always been an only child, a lonely child; she was used to it. The prospect of a couple of years in Japan hadn't seemed all that daunting because what difference did it make, after all?

He told her about himself. Second generation American-born Chinese, too free-spirited for the rigidly traditional family into which he'd been born, too shy to face the world without the shield of a book. No girlfriends; the girls he'd liked had been more interested in jocks and red-blooded rich boys. Never brave enough to venture far from home, the internet had become his realm, and in it he thrived. He was a Big Name Fan in certain circles, known for his biting wit and brutal honesty. The prolif had barely slowed him down.

She worried about what might happen as the clandestine exchanges continued, but never mentioned her fears to him. She'd begun to enjoy herself too much; the "incoming mail" chime was enough to make her heart race with excitement. She had to force herself out for her daily runs.

It helped that the more they talked, the more reliable the relaying became. Pretty soon messages were going through after only two or three tries, and not bouncing at all.

#

IRC session start: Sun? MarEMBJune datetime error

*** marguille sets mode: +o TwenWen Conty Helen sappjuice

> Log set and active! TwenWen logging!

<marguille> dunno why you're logging, twen. it's judst a chat.

<TwenWen> Not just a chat. MadHadder and KT's memorial service.

* Conty sighs.

* Helen observes a moment of silence.

<marguille> ditto. y'know…I herd more spec the other day.

* Conty groans.

* Helen sighs.

* TwenWen waits for Marguille's spec…and waits…and waits.

<TwenWen> ...lagged?

<marguille> this one sounds like it's from the eggheads who did this. here it is: decoherence. when things in a quantum state are coupled to thuings outside that state, both systems collapse. no lag, 2-finger-typing lots, sorry.

<Conty> Yeah, that's egghead all right

* Helen wishes she had a nickel for every egghead spec…but where would she put them all?

*** sappjuice changes topic to "The Egghead Pyramid Scheme!"

* TwenWen giggles.

<marguille> seriously…you heard abut HafCafLatay?

<marguille> she got email from her MOM.

<marguille> as soon as she read it...poof. none of her blog-friends ever heard fm her again.

<Conty> WTF does that have to do with incoherence???

* TwenWen says, "DEcoherence. And I can use other big words, like 'marmalade'."

<Conty> Whatever. Still WTF

<sappjuice> There's spec that *we're* in a quantum state, y'know, each of us. Endless partial variations on the same world, same time...

<Conty> What, so if we ever have contact with somebody in another reality, we'll disappear?

* marguille is typing.

<marguille> the eggheds say it matters if the connection is strong or weajk. the stronger the coupling, the faster the collapse. weak couplings last a long time, maybe even stabilize. vbut with really strong couplings the collapse is nearly instant.

<TwenWen> Ooh, coupling! Wink wink nudge nudge say no more.

<TwenWen> Seriously...you're saying coupling = personal ties? Coupling *to other people*?

<marguille> yep. we're already weakly connected, or we wouldn't be able to talk like this. but strong connections are emotional. HafCaf found her mom and...silence. Both of 'em.

<marguille> sucks, don't it.

<Conty> Ohdamn.

<sappjuice> Huh?

<Conty> I forgot, rollover's about to

*** Conty has been disconnected.

<marguille> Fuck.

<TwenWen> Um, I rollover in 10 mins.

<sappjuice> Maybe we should cut this short, then. Nice seeing everybody face to face, so to speak.

* Helen agrees.

* marguille sighs and waves.

<marguille> back to blogdom then. Toodles.

*** marguille has logged off.

*** TwenWen has stopped logging.

*** TwenWen has logged off.

<sappjuice> So it's just you and me. Wanna email?

<Helen> Yeah.

*** Helen has logged off.

*** sappjuice has logged off.

Session Close: Mon? Time? Deeeeeeeechgkl#@ ^^^^

#

Just spec, she told herself over the next few days. Too many people had expected a more dramatic apocalypse; now they cried wolf at every shadow. Some of their theories sounded right, but most were cockamamie —like Guille's implication that friendship, family, *love,* could be the reason that some people just disappeared. That would mean the only people still alive across the proliferated realities were those whose ties to the world had been weak from the beginning.

Those who'd lived alone. Those who'd been socially isolated. Not the completely disconnected ones; people without 'net access would've gone stark raving within days after the prolif. But the loosely-connected ones, who interacted with others only when they had to, or through a screen. Those who'd maintained just enough connection to keep them sane, then. Just enough connection to keep them alive, now.

Just spec, she thought again as the alarm clock buzzed. She hadn't slept in two rollovers. *Not me.*

New habit. She sat up and reached over to her laptop, which rested on a low table beside the futon, and tapped its touchpad to wake it up. It chimed as the screen lit; she had mail.

"Helen,

I know this is risky, stupid, cheesy, whatever. But I can't help myself. I've never met you and never will, but...some things you can feel no matter what. They used to say this was all just pheromones, but that's crap. I've never smelled you and I only have my imagination to tell me what you look like. But I have to say this because it's true.

I love you.

I wish oh shit I didn't believe it but it's true"

No sig. Not even a period at the end of the last sentence. He'd had enough time to send, but not to finish first.

Not me, her mind whispered, *and not him. Please, not him.*

And as the walls of her tiny apartment began to warp and the barren landscape beyond her window vanished, she had time to click on the bookmark for her blog's "update" form and type a single line.

"The way out, or the end? Sapp's gone to see. I'm going too."

She hit "post" as reality folded into silence.

Poems

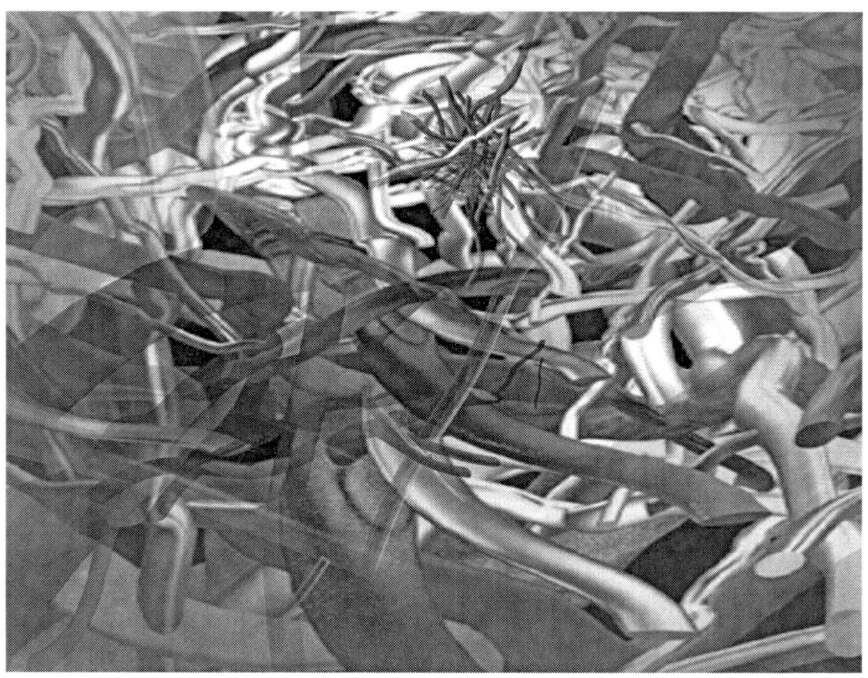

"Superstrings 09-03-05 set 32" by Félix Sorondo

Poetry and String Theory

A conversation between Cecilia Vicuña and James O'Hern

J: For the last few months I've been reading about string theory while you worked on your *Oxford Anthology of Latin American Poetry.* We've had a running dialogue on how these two themes seem entangled. Yesterday, you said maybe it's because we're talking about the nature of language. Of course, no wonder these two subjects feed off one another. Poetry and string theory are both languages that deal with the mystery, the space between the lines.

C: I just found these two lines by an anonymous Nahua poet[1]: *"Todos hemos de jugar patolli/tenemos que ir al lugar del misterio."* We should all play *patolli*[2]/we must enter the place of mystery. That undefined space for me relates to the uncertainty principle. You were just telling me about the Copenhagen Interpretation when science first acknowledged that reality can never be fully known. I think that's the connection with poetry, the point where science and poetry begin their dialogue.

J: But the two disciplines approach uncertainty from opposite sides. Science, seeking a "theory of everything," hits a wall of ambiguity at both ends of the universe. At the big end, it runs into black holes and, at the small end, quantum uncertainty. For science, ambiguity is treated as an enemy while, for poetry, it is the essence.

C: Yes, poetry dwells in the borderland between the known and the unknown where there is so much conflicting energy that all you can do is to acknowledge it by staying in the paradox.

J: In science, if you can't nail it down, it doesn't count. So here we have a clash of worldviews being played out on the field of language with ambiguity as the protagonist.

C: For me, this clash of worldviews is happening both within us, in society itself, and in the poem. You were just telling me about a reading of *The Aeneid* by Virgil. You found yourself in the poem knowing that the gods were there with you as you heard them speak. In other words, their discourse was happening in the moment.

J: Like I time-sliced into the age of Troy. I blew by the constraints of the poem that wanted to hold me outside. I felt their presence like sensing a wild animal in the dark. Then, I got angry with Virgil for selling out to Augustus, for legitimizing violence and power. I wanted to get behind the story, to know what the gods had to say.

C: Last night on the Bill Moyers show, Maxine Hong Kingston was saying that the shape of the poem or the story has the same energy as the life force, so the poem is providing the way in. It is the record of that appearance. As Hölderlin and Heidegger said: being is appearing, standing on its own. In other words, ambiguity is the site where the transformation, the time-slice occurs.

J: Maybe that's my anger. Virgil fills in the blanks for me, not allowing me the space. While hearing the poem it is hard to come up for air. He fills the emptiness with fear and this puts our lizard brain in charge. This is an abuse of poetic powers. He could either open us up to the greater potential or use the deep channels of fear to coerce us into abdicating our own powers.

C: Yes, the blank space lets us be in the not-knowing where ambiguity is still unfolding.

J: But here the heroic worldview gets in the way. *The Aeneid* is at the heart of it. This legacy of violence is still behind modern society. Modern science, like Augustus, needs a Virgil to quell tribal violence and put the empire of science back in charge. The fiefdoms of string theory have gotten way out of hand, embracing mystery instead of mastering it. Many leading physicists see this as bad science, or not science at all. The heroic tradition drives science and wants to kill ambiguity like our "advanced" societies kill the indigenous; the indigenous being the string theorists of history.

C: You once said that truth can be 7 different things and that fear constrains the possibilities to just one. It erases the potential represented by the undefined. In the same way, a narrative that makes things "all clear"—kills poetry. Poetry needs the darkness within.

J: String theory has been around for almost 40 years. It has added life wherever it touches because it validates the creative role of uncertainty. The arts, philosophy, linguistics, and poetry are permeated with its influence. Traditional science takes little notice while string theorists fully participate in this huge paradigm shift. In the meantime, ambiguities keep cropping up in science as it continues its quest for the "theory of everything." What is usually overlooked by science is that even mathematics, the gold standard of precision, is ambiguous at its core. As Gödel said: "numbers, numbers even, are forever beyond human reason."[3]

C: Ambiguity is the DNA of life, the two strands, the union of opposites we call the mystery. Language functions by incorporating conflict, opposition, and ambiguity, and poetry is where language observes itself.

J: This is where consciousness comes in. Science still wants to define consciousness as a personal attribute rather than a fundamental force of nature where we are part of it as much as it is part of us. Science is spooked by the power of the "observer" in quantum experiments, where his mere presence affects the outcome. And it is this reciprocal exchange that connects language, consciousness and quanta.

C: Yes, and this is the main contribution of the indigenous cultures of the world, the awareness of reciprocity. In the Andes people say, "life is *ayni*" (reciprocity).

J: The recent acknowledgement in quantum physics that reality may not exist outside the "observer" implies that we are responsible for creation. I believe the creative act is not only the writing of the poem but occurs each time the reader participates in the ritual. Each reading of a poem is a measurement as one measures the movement of the stars.

C: Responsibility is the ability to respond. In the Andean mythopoeic universe, if you do not accept responsibility for your part in the cosmic exchange, you cannot die properly, you become a *condenado,* a condemned soul that wanders between dimensions, tormented and tormenting your relatives and loved ones because you could not let go of this world.

J: The act of measurement has been sacred since time began. It is the function of the astrologer, alchemist, and oracle—the poet and the shaman. In quantum physics, the measurer is the "observer."

C: Measuring, then is the creative act, our choice in the instant.

Juan Sánchez Peláez[4] says:

> *"la obra humana es el instante"*

the work of humanness is the instant

and Luis Cardoza y Aragón[5] says:

"la poesía es la única prueba concreta de la existencia del hombre"
poetry is the only concrete proof of the existence of mankind

Notes

1. From *Cantares Mexicanos* quoted by Miguel León Portilla in *Los Antiguos Mexicanos,* p. 185.
2. Aztec/Mayan board game.
3. Janna Levin, *A Madman Dreams of Turing Machines,* p. 203.
4. Venezuelan poet, reprinted from *Antología de la Poesía Hispano-Americana Moderna II.*
5. Guatemalan poet, source unknown.

Cosmic Gambol

Colette Inez

with the persistence of strings
afloat in space-time, I walk
the Planck-length, blind
to its millionth of a billionth…
centimeter. It's a narrow path
to find my boson mate, elusive Higgs,
as I desire my *colettions*
to loop and loop in wavicles
of joy that make me matter.

String Theory

Dave Morrison

Thousands of lives are being
spent trying to construct an
Explanation of Everything, six

extra dimensions, parallel universes,
the whole overloaded wagon held
together with tiny bits of string.

String?

Why not tiny jacks, or tiny crescent
wrenches or tiny dandelions, or tiny
charm bracelets, or tiny umbrellas, or

tiny teacups, or tiny disco balls, or
tiny rubber bands, paper clips?
Our Universe is as it is because of

tiny hoops of twine? This
is what you get from too much
coffee and a 'what if' —

When my friends and I considered
such possibilities behind the
cafeteria with a nickel bag of

lame weed, it did not occur
to us that we were budding physicists;
we were burnouts
trying to get comfortable with

mystery.

Life x 10^{-33}

Joseph Radke

– for JP

No one denies the unseen –
the rush of guardian wings,
the whirl of charmed quarks.
And who could doubt
the supersymmetry of the universe.
We know love and the beloved, forces

and things that matter.
No, we can't renounce
the invisible, the fluid foundation
of the solidly seen. We can
only imagine and speak
in shrinking untruths.

So it seems God's faithful
angels have tired of dancing
on pinheads to the horns
of creation and are wracking
their branes to tune
the superstrings
of their harps' fundamentals.

A Mapped Route to the Island of __________

Kathleen M. Heideman

Was there fog? Can I blame my navigation errors
on an ordinary layer of interference — say the moon
was luminous at first, large as the eye of a lighthouse,
only clouds came later? Well, then, yes.

It is possible to paddle purposefully for hours
and still miss the shore of Romance. ((So close — almost!))
At least that's what a friend said, explaining how he kayaked
out to the islands of the Sleeping Giant, and nearly missed

— not by much, perhaps, but aim is not enough, such nights!
So often my buoyant little life has slipped a single mile
too far west of the rocky point that is a mooring place,
or will become so, soon, for someone luckier at steering.

Still, an island is a looming fact, visible from a distance.
Igneous, the basalt sides form lines of forced intrusion;
they weather slower than the surrounding landscape.
How clearly an island reveals the world's inner nature:

old crust floating on a deeper sea. Notice me veering
to avoid the thing I'm aiming to say, even as I break my own
silence to say so? If only I were better at calculus,
my navigation might improve, my poems might steer

more easily toward the certain landmass of Conclusions.
I don't mean simple arithmetic. Complicated as my
checkbook gets, it's not enough to track down dollars,
cents. There's something deeper I want: higher math,

a happy accident, what happens when you draw an island
and a boat and braid your strings into a rope — suddenly,
the negative space between those shapes becomes activated,
a gap grows into an thing with its own meaning: Hope,

liquid, mappable. Sometimes, summer nights,
I take a bottle of red wine to the garden, and a notebook, and
this pen. I write myself into existence. More than once,
moon's beam illuminates how I've scribbled the same equation

down the page, solving for *love,* an untutored stutter:
1+1=2. (1+1)==(1+1).
How many possible routes, how many maps rolled into tubes,
tubes pulled taught into strings, folding into points?

Fine, I'm expecting too much from romance.
Certain molecules, they bond whenever they meet…H_2O, you know.
Other times they turn caustic, discolor their container, explode.
I've had lovers like that. Or worse: inert; there's no reaction,

1x1=1. Recent studies in the U.S. suggest millions of respondents
are "very satisfied" with basic math. Fog arrives while paddling.
You think you know my answer now, don't you? We are hoping

for the existence of things quite difficult to prove. Dimensions,
pulled like taffy strings. In Chicago, scientists are launching accelerated
particles north, through the bedrock of Minnesota: through dolomite,
Laurentian greenstone, Gabbro basalt, Vermilion hematite deposits…

The aim is a vault of iron plates in the labyrinth of an old iron mine,
a lab deep underground. Down they go, later, to check for a proof
which arrives in the form of miniscule holes burned in iron,
a scatter-pattern of subatomic dots. The M.I.N.O.S. Project, they name it.

And what was myth becomes statistic —our aim is refined; particles
hit the target with regularity, but we are looking to unravel
something more elusive; a hint, a route by which 1 reaching 1
reveals the stray mark of a degrading neutrino…

*

Whenever a conversation shoots right over my head, I try to picture
something simple — a million dots of color, a pointillism painting by
Georges Seurat. If you stand too close, the gallery alarm goes off,
and the dots are just another crowded Friday bar, enisled points.

But squint a bit, and the dots vibrate to form a picnic scene, *"Sunday Afternoon on the Island of La Grande Jatte."* Pure color, a landscape of spring umbrellas,
fancy hats, strolling couples, narrow boats rowing along the Seine.
Specks of paint, dots placed very close to the next, shimmer like cobwebs

& talk until their stories blur together. Seen from far enough away,
something else is always glimpsed — astronauts who reach the moon
report how small our continents seem, small islands of the earth, dots.
And that string of islands, northeast of Thunder Bay?

They form the outline of a resting man: Sleeping Giant, Manibozho.
It is possible to paddle out there, land, and fall asleep with your head
on the island's stone thigh. Better paddlers than me have studied mythology
but miss the island, lose their way, listening between each stroke...

Scared. Trying to calculate horizon, the line strung between water and night.
The route is mapped, a dotted line. Between the Isle of the Heart
and the Isle of the Head, the lake runs wide and deep. In theory, it is possible
to push straight through, unknowingly. *As you are doing, yes. Like so.*

String Theory

Bruce Holland Rogers

1. The Eleventh Dimension

No one had direct experience of the extra dimensions, so learning about them at age six was difficult even for genetically enhanced superchildren. Child-friendly names were devised. This helped. Children learned that the eleven dimensions were Length, Width, Height, Time, Happy, Sneezy, Dopey, Grumpy, Sleepy, Doc, and…Even with the new names, one dimension was hard to remember. Even for genetically enhanced superchildren, the universe was not without mystery.

2. Recipe for a Theory of Everything

Start with a figurine of turtles stacked one on top of another and an excellent hammer. Smash the turtles. Smash the pieces, and keep smashing. Pound the dust into atoms. Smash the atoms into protons. Keep smashing down to quarks and gluons. You're close to the theory of everything. Pound everything into strings. Keep pounding. After strings, turtles. Pound, pound, pound. Smaller and smaller turtles, all the way down.

3. But Maybe She Just Couldn't Knit

Wanda had been about to defend her superstring dissertation when the universe gave its answer. Broken strings littered the floors of physics departments everywhere. "But they were so pretty!" she cried.

She drank.

Later, she picked herself up. She went to AA meetings. She spun her old strings into yarn, knitted the yarn into a sweater and wore it to a meeting. Everyone who saw it started drinking again.

7 Proofs of the Existence of Quarks

Michelle Morgan

Proof #1

I have been to the other side so you know what I say here is true.
Shopping at a vacant Meijer in Columbus during a Buckeyes game,
I knew that all of that talk about parallel worlds was no joke.

Proof #2

I have also been stuck in interstellar traffic behind a guy
in an orange RV with a bumper sticker that read,
In the event of a wormhole opening up, this vehicle will immediately vanish.

His girlfriend exhibited all of the classic hallmarks of supersymmetry.
She was very thin. I know this because a tear in the time-space fabric
caused my car to suddenly ram into their fender.

She claimed a case of whiplash, said her intestines were all in a loop.
Cosmic policemen took photos, but I lost them.
The pictures were very small.

Proof #3

At the measurement of 10^{-35} I have been known to jump across all 26
dimensions.
I am so tiny science has not figured out a way to define me.
I am beyond the means of modern telescopes. I find that reassuring.

Proof #4

At one point or another, I have been accused of *stringing-people-along*.
The only proof of this phenomenon is the fact that I have no idea what
they are talking about.
If you find yourself meeting up with me somewhere, please explain.

Proof #5

I have seen the effects of string cheese on my son.
He positively quivers.
It is really something.

Proof #6

Once I was on my wireless telephone and I could hear *someone else's conversation.*
It was the middle of the night, and I was all alone.
Larry, let your mother know that Jack will pick her up for her dentist appointment at nine.

Proof #7

There are women in this universe who actually wear a size 0.
Apparently many, many women, since all the chicest clothes only come in that size.
This is the culmination of everything I could ever hope to attain.

Building Blocks

Robert Borski

Morons are not elementary particles,
birds do not quark, and half-
dead cats fail to constitute roadkill.

On the other hand
silly string may underpin much
of the universe, and in
the toychest of infinities
both larger and smaller sets can be
found side by side.

God as a boy must have been
a strange child, if not actually gifted.

Ghazal Proof

Sandy Beck

"String Theory" is what Theoretical Physicists are now trying to prove.
Known as "The Theory of Everything"—how can it be possible to prove?

My mother feared my father's medieval sword collection. He taught me to fence
when I was ten, warned me: *someday you might find you'll have a lot to prove.*

Male Bowerbirds compete for females by gathering the brightest trinkets
with which to decorate their nests. Resourcefulness is what they want to prove.

Before young Isaac played his violin for a cohort of white-haired musicians,
he first decided: my genius is a given—not anything I will ever need to prove.

Evelyn relied on the effects of Cocaine and Opium to enhance her orgasms.
To her they were not evidence of womanhood but skills she had to prove.

Beckett weighed one hundred pounds and spoke in a high voice. He dated
ten girls in three months. Do you think he had something to prove?

My friend Gus got out of his physics exam. Disguised, he kidnapped the teacher
and put her on a plane to Paris. To this day, there is still nothing to prove.

String Theory

Elaine Terranova

What's more basic than string?
Cut string, you still have a string.
Like the Brahmin's idea of unity,
pouring milk into milk, which gives you
milk.
        Really, that's the opposite.
 It's adding, not taking away.

Strings are loops of energy, vibrating.

String, that you bite off with your teeth?

String that's supersymmetric particles
passing through your house and
your body.
                Knocking things off shelves?

Causing internal changes. They jump
the way shadows jump. They jump like birds
and startled animals.
                                Yes, it's true
I am not always myself but sometimes
a person I could become.

One way to think of it: Begin with a line.
Remove the middle third. Remove
the middle third of the segments. Do it again
and so on. Only a dust of points remains.

I once drew a picture gluing on string
to stand for the rain. I don't like rain.
It seems to be reaching out for you.

Remember that all motion is trying
to be perfected, to be still.

Yes, I imagine
the divine patience of the poor,

waiting in line for buses and bathrooms.
And the sea at the horizon.

The line of the sea
is not really a straight line
because you can never unravel the shore.

The snake I saw that day snapped like a line,
like a string or a clothesline. Yellow and black,
it snapped through the leaves. Like them,
yet separate. Low like them, the height
of a shoe. Crinkling like them,
the snake and leaves, dual to one another.

Physics For Dummies In 3D

Diane Shipley DeCillis

She couldn't fall asleep, had watched a program on PBS about **string theories**, and **membranes**, (**branes** for short) and **parallel universes**, **gravity—gravity vs. electromagnetic forces**, **atom smashing**—alright it wasn't as if she understood **physics**, additional **dimensions** beyond the three that she, and most everyone are familiar with: **up and down, side to side, back and forth**—really, what other **dimensions** do we need, she has trouble fitting all that **motion** in her day as it is, so many things to do... sometimes she wished she could slow down, **molecularly**, maybe live in the country or on an island, but that could just be boring as hell, might *BE* hell, no visceral stimulation, the adrenaline she craves, seems addicted to, and then she began to think of when she'd been agoraphobic and how even *that* might relate to these **string** etc., **theories—dimensions of time and space—up and down** was fine, **side to side**, okay, **back and forth** was problematic; her **dimensions** had shrunk to two and yeah maybe she could have used a few more, but there was a time when even a photograph of a lone figure paddling across a river in a red canoe against a Vermont blue sky with crimson and gold leaves, like daytime **stars**, caused anxiety; and if those **physics** professors on PBS had been watching they might have associated the **ripples** in the river with the **reverberations** of her nerves and might have asked her what she was afraid of and she might have said, "I'm afraid of heights, only **horizontal** instead of **vertical**," and those brainy men, some with skinny bodies because they eat, sleep, and drink **science**, some with fat lips and fat earlobes, (who knows why), some with white hair that seems to be a **conductor** for **electromagnetic impulses** might have thought these fears were alien, of **another universe**, but one she understood as a **universe** which began when a father left home before his little girl grew up, where the **pull of gravity** is the tug of a heart, where **time and space,** maybe even **air** itself seem **infinitesimal**, and that might explain how she could feel alone and question things that are beyond her reach, at night when she is trying to fall asleep in the dark—the **stars** too faint to notice.

Up & Down

She couldn't fall asleep—
string theories, membranes
vs. electromagnetic forces
physics, dimensions
back and forth
that motion in her day
she could slow down
on an island but that
stimulation
she began to think about
these string etc., theories
side to side, okay
had shrunk to two maybe
when a photograph of
a Vermont blue sky
caused anxiety
they might have associated
might have questioned
I'm afraid of heights
some with skinny bodies
and fat earlobes
like a conductor
these fears were alien
as a universe which began
where the pull of gravity
maybe even air itself
says she could feel alone
trying to fall asleep

Side To Side

had watched
parallel universes,
atom-smashing—all right
beyond the three
what other dimensions
so many things to do
molecularly,
could just be boring as hell,
the adrenaline she craves,
agoraphobia
dimensions of time and space
back and forth—problematic
she could have used a few more,
a lone figure paddling across a river
crimson and gold leaves
if those professors on PBS, saw
the ripples in the river
what she was afraid of
only horizontal instead of vertical
they eat, sleep, drink science
(who knows why),
for electromagnetic impulses
of another universe
when a father left home
seems visceral,
seems infinitesimal
question things that are beyond
in the dark—

Back & Forth

a program on PBS about
gravity—gravity
it wasn't as if she grasped
up and down, side to side,
she has trouble fitting all
sometimes she wished
maybe to live in the country
might *be* hell, no jarring
seems addicted to
how *that* might relate to
up and down, fine
her dimensions
but there was a time
red canoe sparkled
like daytime stars,
had been watching
reverberations of nerves
she might have said
and those brainy men,
some with fat lips
some with white hair
might have thought
but with depth she understood
before his child grew up,
where time and space,
might explain
her reach, at night
the stars too faint to notice.

Hook of a Number

Lauren Gunderson

the hook of a number
sixes and twos
clutch quanta, catch thread
make bed of messy theory of everything.
but the numbers will tell.

the hook of some number
the hollows of eights
the dots and slip knots
of abstracted things hand made into sense
will tell.
the ordering will come.
the tight weave will expose its structure
and dimensions will blink in our eyes.

The look of a calm place can already be known
as tiny treacheries of space-time flux,
an elegant turbulence whipped up and down
made flatly 3-D only by our largeness.

What next of these maths?
What proof, what sell?
the word topology chews itself.
branes fly sting-ray-smooth
past waiting wonderers.
how long a wait?
angels on pinheads come to mind.

The hook of a number
    akimbo seven
god-mouth ought
thumbing nine
spent up learning the twist
of these strings.
since we can't kiss them our selves,
the numbers will tell.

String Theory

Cherryl E. Garner

Matter and its opposite
hiccup, gulped
in the big lipped,
ink-black hole.

We'll visit, then be gone
before we letter write,
flat in singularity
with our undone strings.

We'll not get note from
God. He'll lay a kind last
hand, poetic pulse,
on us, clap the erasers

until we're less than
puff, slit by
passing planets
into branes.

We'll simple carbon pitch,
then diamond white.
Planes, dumb, will twist
done night with wail.

Gravity Spool

a renku scifaiku string

Mary Margaret Serpento, ushi, oino sakai, Deborah P. Kolodji, assu, and Lucinda Borkenhagen

gravity spool
a loosened thread
draws her in — mms

a remarkable proof
when the apron balloons — u

strongly tied
a galaxy held
in space-time — os

equations knit
a symmetry of nature — dpk

short dimensions tickle
loose feet scuff charge
but to what end — assu

lab coats crackle in the darkness
a spark as their noses touch — u

shadows
inside Moon craters
my hiding place — lb

in line-of-sight
palimpsest spectrograph — mms

missing world lines—
a corpse flower bloom
no one sees — dpk

necklace of worlds
galaxy on the head of a pin — assu

uncontrolled torque
colliding galaxies
twist and implode os

their godlike powers focus
to housebreak the new pet u

Finnegans Elegant Universe or *The Elegant Wake*

Being A *Quasi* Random Sampling/Mash-up Poem: *Finnegans Wake* + *The Elegant Universe*

Michael Ricciardi

Like a double slit experiment / like particle wave duality / we rely on Leibniz equivalency / to maintain the smallest and largest / of radii / of sizes of things / these superstrings / of thousand fold form / in queerest symmetry / of manifold geometry

Hark! I Hear Dimension's Lament:

Once / all we needed was 3 / then 4 / ('tis true!) / but now it seems: 11 / and no less will do / to unify the fields / the quantum / and the continuum / uncertainty / and relativity / we demand higher dimensionality!
Once / all we needed / was Newton – Einstein / but Now / its Kaluza – Klein / and Calabi-Yau...

Fallright-ie then! I'm picking up super vibrations...from my GUT to my TOE...

Let me finger their eurhythmytic. And you'll see I'm self-thought...a pushpull, qq: quiescence, pp: {with extravent intervolve coupling}

Oh Know! The Wave of Un-Certainty Cometh!

We can know *q* but not *p* / how can we see / without interference? / some quantum incoherence / and the wise-fool thus spake:

Thence must any whatyoulike in the power of empthood be either greater ***Than or less thaN*** *the unitate we have in one...*

Impassable!

Living out all possible futures / in the universe next door / or the next door after that / there's plenty more / a housing boom / it seems / between big bang / and Planck mean-time / and the wise fool chimed on:

...hence shall the vectorious ready-eyes of evertwo circumflicksrent searclhers never film in the ellipsities of the their gyribouts those fickers which are returnally reprodictive of themselves...

Quarrellary! And a Feynman he is!

Here we are / chasing the Horizon (problem) / and waiting for particles to appear / in machines made of particles / made to interfere / You can never! / You lie! / cried the theorist / and smashed on / to verify the absolute minimum something / or its partnering / a shadow of a spasm-ing string / of what else to call it / but god?

But lo, the wise-fool thus spake:

The logos of somewome to that base anything, when most characteristically mantissa minus, comes to nullum in the endth...

Ah, the Reductio! Deny!

There be nothing beyond / or smaller still / we but wait on higher planes / or fields / seeking confirmation / of the gauge / in the infinitesimal fact of matter

What bravura! What pity!

And even if we fail / we fail elegantly / in our calculus dreams / clinging to a Planck length / adrift upon on a Dirac Sea / flatland only a memory / buffeted by waves / of probability / 'til perchance behold: / The Symmetry Duality! / (or some hidden dimensionality) / and sets us floating / passed Atom & E/*ve*'S / on non-Riemannian geometries

astronauts: three excitation modes

David Hurst

(after Ted Berrigan)

i

they can't admit in the penetrating
klieg glare of television lights
their natural urge to suckle.
twisting in the expanse of space
tethered to a silver colonial phallus
of coldwar passion, they must know
only their orbital speed keeps them
from falling into recurring dreams
of a great mottled blue areola.
they must glance at it and look quickly
away, moistening their lips with guilty tongues
fumbling about in thick sheaths.
surely they wake each day as i do, stiff,
the memory of her swaying breasts just fading.

ii

they can't admit to penetrating
their moistened lips with guilty tongues.
tethered to a silver ovary, twisting
in a great womb of coldwar passion,
only the memory of a mottled blue areola
prevents their recurring dreams
from glancing into the klieg glare
of television lights. their orbital speed
keeps their thick sheaths from fumbling
into the natural urge to suckle;
they must know, as they glance
and look quickly away,
i wake each day as they do, stiff,
the fading sway of her breasts just a memory.

iii
they can't admit to tethering
the great just passion of a coldwar
colonial phallus to the
mottled blue areola of space.
only the memory of guilty tongues
prevents the silver expanse of television
from fumbling about in thick sheaths;
they must know their natural urge to suckle
keeps the klieg glare of lights
glancing quickly away. their moist lips
dream of the orbital speed of ovaries
and the recurring penetration of memory.
surely, as stiff as i each day in the fading
of the sway of her breasts, they must wake.

String Theory Sutra

Brenda Hillman

There are so many types of
"personal" in poetry. The "I" is   a needle some find useful, though
the thread, of course, is shadow.
In writing of experience or beauty,   a cloth emerges as if made
from a twin existence. It's July
4: air is full of mistaken   stars & the wiggly half-zeroes stripes
make when folded into fabric meant
never to touch ground ever again—   the curved cloth of Sleeping Beauty
around 1310, decades after the spinning
wheel gathered stray fibers in a   whir of spindles before the swath
of the industrial revolution, & by
1769 a thread stiff enough for   the warp of cotton fabric from
the spinning frame, the spinning jenny,
the spinning "mule" or muslin wheel,   which wasn't patented. By *its*, I
mean *our*, for we would become
what we made. String theory posits   no events when it isn't a
metaphor; donut twists in matter—10
to the minus 33 cm—its   inverted fragments like Bay Area poetry;
numbers start the world for grown-ups,
& wobbly fibers, coaxed from eternity,   are stuffed into stems of dates
like today so the way people
are proud of their flag can   enter the pipes of a 4.
Blithe astonishment in the holiday music
over the picnickers: a man waves   from his spandex biking outfit, cloth
that both has & hasn't lost
its nature. Unexpected folds are part   of form where our park is
kissed by eucalyptus insect noises ^^z-
z~~> *crr*, making that for you.   Flag cloth has this singing quality.
Airline pilots wear wool blend flag
ties from Target to protect their   hearts. Women, making weavings of
unicorns in castles, hummed as they sewed
spiral horns with thread so real   it floated; these artists were visited
by figures in beyond-type garments so
they could ask how to live.   It's all a kind of seam.
Flying shuttles, 1733, made weaving like
experience, full of terrible accidents &   progress. Flags for the present war
were made in countries we bombed
in the last war. By *we*   you mean *they*. By *you* it
means *the poem*. By *it*, I
mean meanings which hang tatters of   dawn's early light in wrinkled sections of
the druid oak with skinny linguistic
branches, Indo-European roots & the   weird particle earth spirits. A voice

came to me in a dream
beyond time: *love, we are your*
shadow thread~~ A little owl
with stereo eyes spoke over my
head. I am a seamstress for
the missing queen. The unicorn can't
hear. It puts its head on
our laps. Fibers, beauty at a
low level, fabric styles, the cottage
industry of thought. Threads inspired this
textile picnic: the satin ponytail holder,
the gauze pads inside Band Aids,
saris, shine of the basketball jersey,
turbans, leis over pink shorts, sports
bras: A young doctor told us
—he's like Chekhov, an atheist believer
in what's here—that sometimes, sitting
with his dying patients, he says,
"God bless you." It seems to
help somewhat. They don't know what
causes delays between strings—by *they*,
I mean the *internet*. Turns out
all forces are similar to gravity.
We searched for meaning ceaselessly. By
we I mean *we*. My sisters
& I worked for the missing
queen: she said: be what you
aren't. A paradox. There are some
revolutions: rips in matter, the bent
nots inside our fabric whirred &
barely mattered any more. Our art
could help take vividness to people
but only if they had food.
No revolution helped the workers, ever,
very long. Tribes were looser
than nations, nations did some good
but not so very always, &
the types of personal in art
turned & turned. Nylon parachutes for
World War II. We shall not flag
nor fail, wrote Churchill. O knight,
tie our scarf on your neck.
Je est un autre wrote Rimbaud
the gun-runner. Over & inner &
code. The unicorn, *c'est moi*. The
rips by which the strings are
tethered to their opposites like concepts
of an art which each example
will undo. We spoke of meanings.
I, it, we, you, he, they
am, is, are sick about America.
Colors forgive flags—red as the
fireskirt of the goddess Asherah, white
as the gravity behind her eye,
blue for the horizon unbuttoned so
the next world can get through.
It's not a choice between art
& life, we know this, but still:
How shall we live ? *O shadow*
thread. After the workers' lockout 1922,
owners cut back sweatshop hours to
44 per week. The slippage between
string & theory makes air seem invented
& perhaps it is, skepticism mixed
with fear that since nothing has
singular purpose, we should not act.

To make reality more bearable for
some besides ourselves? There's a moment
in Southey's journal when a tomb
is opened & the "glow-beast" exits—
right when the flying shuttle has
revolutionized their work—by *their* I
mean *our* – & cut costs by
half. So lines are cut to
continue them & if you do
help the others, don't tell. String theory
posits symmetry or weight. My country
'tis of installing provisional governments.
Why was love the meaning thread.
Textiles give off tiny singing no
matter what: washable rayon, airport
carpets, checked flannel smocks of nurses,
caps, pillowcases, prom sashes, & barbecue
aprons with insignias or socks people
wear before/during sexual thrills after
dark subtitled Berkeley movies next to
t-shirts worn by crowds in raincoats.
Human fabric is dragged out, being
is sewn with terror or awe
which is also joy. Einstein called mystery
of existence "the fundamental emotion."
You were unraveled in childhood till
you were everything. By *everything* I mean
everything. The unicorn puts its head
on your lap; from there it
sees the blurry edge. How am
I so unreal & yet my
thread is real it asks sleepily ~

Raise It Up in the Mind of Me: One Poem, Eleven[1] Footnotes[2]

Jeff P. Jones

Women have small taste for the sea.[3] – Melville[4]

But there's that overnight ferry[5] between Stockholm and Helsinki
she took by herself. Tucked into a sleeping bag against the cold wind,
the hum of giant engines at her back, she saw a shattered burst of light,[6]

red and white and gold[7] crawling across the night sky. She pulls
another shot of espresso, glances at the gray snow[8]
falling and the mediocrity-worshiping world. If she moved

further west, even Portland or Seattle, she would be closer.
Snow in this land-locked place brings a special anxiety.[9]
Each layer a covering. Unstoppable. Inevitable.

It's beginning to stick, and the boy[10] who makes love to her
will soon step through the door and knock snow from his coat.
He'll look up and with a smile breaking his face say, Hey.[11]

[1] String theory posits up to eleven dimensions. This goes beyond high school trigonometry. At least one of these dimensions blurs at the speed of a plucked guitar string, becomes uniformly invisible and able to inhabit more space than it ought.
[2] This poem was originally titled "The Special Anxiety Brought By Snow."
[3] It takes twelve seconds to handwrite this line.
[4] Another three for this one.
[5] Screen is black. We HEAR a woman's voice. We can sense her compassion, her deep emotional reservoir.
[6] In southern Colorado there are rocks that catch fire when struck by lightning.
[7] Golden bits of titanium carved from the shoulder of a Russian satellite.
[8] Picture here a duck's body, egg-white, loose as a fallen tree branch, somersaulting through the water.
[9] Me at my most neurotic: Please don't think of me as a ghost. I myself am afraid of ghosts because they don't like me, they laugh at me, they hate me, they think I'm stupid, they see right through me.
[10] Dale Evers could have been a glam-cowboy success in the tradition of Roy Rogers and Gene Autry. On one shoot, the whole cast was bused out to a desert location. Dale was miffed. Yet a young Indian woman caught his eye and he chatted her up. She'll make this week worth my while, he thought.
[11] Some people believe that sound never dies but continues to reverberate. A conversation from twenty years ago might still be in this room, bounding off the walls. I want to believe this.

Longing

Heather Holliger

towards a physics of meaning

Interrogative

I am asking about a cosmos
of numerical delineation:
the physics of matter, of space-time curves.
I am asking about perception:
beyond reds and blues,
the probability
of more than four dimensions.
The unimaginable—pulsars
and photons, a single light-year.
Expansionary fabrics—
space   canvas   consciousness.

Theorem

String Theory (also known in physics as a "theory of everything")

↓

Matter = A composition of strings,
two-dimensional, infinitesimal,
smaller even than quarks.
Loops of vibrating strings—frequencies
configuring space and time
numerating
our senses   our skins   our desires.

Harmonics

I ache for resonance, luminosity—
perennial moments, glistening leaves.
For the sensuality of green—its taste,
touch; flora as metaphor, as sentience.

I bend towards the rhythms, affinities
of touch—a wisp of hair, red sepal of lip;
the scent of garlic, of sugars on the skin,
the music urging embodying desire.

It's the liqueur, the breadth, of which I write—
transmutations of light; of what is visual,
and unknown—domes of indigo, and blacknesses,
constellations, vespers, the touch of hands.

My fingers crush into/out of meaning—
Liebesträum No. 3 in A flat Major; the intensity
of touch, a maddening locomotion,
imperatives imbuing shimmering the strings.

I long for glimmers—sunlight, shadow,
autumn's reflection in pools of blue.
For waterfalls—time, cloves, recesses,
red petals, touch; beyond
spaciousness, loneliness.

The Theory Of Everything

Linda Nemec Foster

> "…a vision of physical reality so at odds with our experience that it defies language." – The Teaching Co.

But what is it? What noun
placed next to what verb
modified by what adverb
holds the secret? I ask my daughter,
the aspiring astrophysicist,
to speak the sentence; and if not
the concise sentence, then the undulating
lines of pure thought that describe
the theory of everything.

Her mother, who froze in high school
geometry, cannot comprehend how
tiny strings vibrating in
a microscopic universe can hold
everything together: from DNA's
double helix to the silky
translucence of a moth's wings
to Bach's *Concerto for Two Violins.*
How it can all be reflected
in eleven dimensions: eleven
parallel universes wrapped
in empty space—a dark energy
of nothing. My one-dimensional
mind boggles as my daughter explains.
But the messy world of an atom's nucleus
(the photons and quarks, the positrons
and muons, the wimps and Higg's Boson)
all blur in my tired head.

She describes a famous physicist's
lecture and I can only imagine
him at the podium with mismatched
socks. Dark blue of sky mistaken
for dark black of night. No use
searching my finite space for a unified

theory when I can hardly recognize
my own daughter as she lives
more and more in her own universe
and leaves my small world behind.

The daughter who waxes and wanes
like the moon; loves me
and pulls away like the tides;
listens to a rock group
called Magnetic Field sing
about the unscientific mess
of love; loses car keys
and forgets to turn off
the stove when the primordial
soup boils down to nothing.
The daughter who as a child
was lost in a Chicago museum
filled with the physics of Magritte;
and as a smaller child noticed
the silica shimmering
in a lake in Nova Scotia
and deemed it diamonds. This
woman who now peers at the stars
in the night sky and sees
just as many diamonds. And
in the morning thinks the warm
air of a January thaw is not
fog, but the broken snow
on fire. The woman who knows
the textbook explanation, yet
wants to believe in the flames.

The daughter who looks at me
with my cosmology of tentative
words, tentative silence and tries
to see the mother: the proof
that experience does defy
all language. And everything,
everything is connected. Whether
we can dare to believe it or not.

Confessions Of Gaver Immer

Susan Zwinger

I am awkward on land, a body thrust forward,
tennis is impossible. In higher realms,
I could never keep up with the terns.
Sheer strength carries me thousands of miles,
north over land, south over sea.

I skulk checkered in shadows.
No one but Grandmother Immer
understands me, she gives me
her heirloom pearls for my elegant neck
to be worn to interviews. A checkered
career stalls out. I hover in a poet's
gloamwater light nesting on floating islands.

My echoing tremolo across dawn lake
freezes people in their tracks with longing.
They spin around to catch sight, but I'm under
swimming the eleven strings
of a space they never will enter,
through eleven curled membranes
they can never imagine.

That Reckon the Corn Before It Ripens

Wendy Vardaman

And if we're wrung and left hung out
to dry? Rolled hems stiffening
in the wind, strung upon a sagging stretch

of line, forgotten underneath a fallen
sky? Staggered poles decline in opposite
directions. Who'd ever have imagined

laundry from these heights, and with
the clash and crash of water over rock?
Whose faded shirts are these and whose patched

knees? What disembodied flock,
blown back and forth but always unaware,
through curled dimensions that like the folded sock,

lie hidden one inside the other,
hidden inside ordinary air.

String Theory

Christine Klocek-Lim

The mind makes pieces of things:
string theory, evolution,
the unpredictable way a tree
falls under lightning.
And I have observed the world
slow before surgery,
seen how even motes of dust
become important in a hospital.
But there is solace in the ordinary.
The universe is littered with stars
and mass. Every night the earth
falls into dark.

And I don't want the day to pass stuttered
with spilled tea, indecisive steps.
The bewildering routine we follow
sends us to bed tired and lost,
frightened of the looming storm
that threatens trees in the front yard.
The physics of living tangles the common,
like the stuff I've jammed
in the kitchen drawer and forgotten,

where there is another universe
filled with old rubber bands,
defunct batteries, and cotton
cord wound into a ball.
But these are not broken items
where a string loops closed.
No. The batteries passed
their energy to a flashlight.
The rubber bands defy gravity,
hold together notes written
in the hospital, in another world
where words controlled the way
our son survived.

Years have passed since then
but the paper says the same thing:
11 am, check the IV for air bubbles.
Amazing how the invisible can explode
a vein, how words written can shift time
from past to present.
It's all relative.
Many strings compact to one
when the universe shrinks into a dot.
The mother of a sick child knows this forever.

When I clean out the drawer, I find
the string's end has frayed into ten strands,
which themselves can be unwound
into twenty-six. Another choice looms
but I'm out of time and loop the string
around the notes. It'll hold for a while.
The past is bound to the mind
in pieces that strike like lightning
despite the storm's end,
despite the way evolution
changes even dust.

Crossing Over

Cleo Fellers Kocol

Central explains string theory
in everyday language. With light/
time/distance balanced by substance,
dismantled DNA strings are
re-connected post transmission.

> *Twice a chimp signed for*
> *a banana after reassembly.*

Dark matter evidently holds a key to
communicating across dimensions, and
moral debates escalate as kids finger
instant understanding, computing
modalities that whiz by like previous
generations' falling stars.

> *I pretend to understand, but*
> *all I care about is maintaining*
> *what we had—a two kid family,*
> *shuttle trips within the orbit.*

I enter the kids' rooms and silence follows.
Lately, Cissie and Buck host meetings
from Over There. Projected faces flat,
visitors stare past me. Side-stepping
holographic images, I leave. The decibel
level rises and grim-faceted laughter follows
light-beam images mind-trolling my thoughts.
Increasing numbers of solid visitors gaze
past me with uncaring recognition, voices
clear, gilded gentility flaking when I question.

> *I only want to protect my kids*
> *from sludge siphoning their*
> *homeland energies, but I appear*
> *retrograde, deficient in the process.*

Today my son's room is empty, his vid screen
down. Four days later—the same. Citizens
Data Bank Responder says his name has been
deleted, no further statistics available.

It's as if he never existed, as if I'd
never taken him from the lab. His
sister's memory board sparkles
messages I cannot decipher. Mine
asks if I want to contact adults
on the other side.

String Theory

Beret Skorpen-Tifft

> "if string theory is correct, the entire world is made of strings!"
> – Alberto Guijoso

I always knew geese were made of string.
Perhaps it was the way they noticed
a new wind carving through corn
or how they gathered like tourists
queuing for a ride, a little unsettled,
somewhat eager.
Their running departure was necessarily
heartbreaking and thrilling.
Oily crows, beige gulls, finches lingered,
like me, and in no time we started to unravel.
We spent the coldest days searching
for our ends. The crows, finders of everything,
found theirs quickly. The rest of us waited,
growing smaller. We might have gone back
to the very beginning if not for the geese,
who came back one day
plump, loud, and whole.

Drama

"Superstrings 26-12-04 set 22" by Félix Sorondo

Strings

Carole Buggé

Cast

JUNE	40's.	Slim, fit, a rock climber, a cosmologist. American.
GEORGE	40's - 50's.	Absent-minded, idealistic, a bit remote. Also a cosmologist. English, upper class.
RORY	40's.	A math whiz, but somewhat childlike and insecure because of his lower middle class background. English. A particle physicist.
ISAAC NEWTON	50's.	Tall and imposing, dressed in 17th century garb. Brilliant and arrogant. English.
MARIE CURIE	40's.	Small, with short dark hair. Dressed in 19th century garb. Quiet, dignified, gracious. French accent.
MAX PLANCK	40's.	Slight, balding, mustache, delicate features. A sweet man, in spite of a certain old-fashioned formality. Prussian accent.

Production note: Though Rory and Planck do play cello and piano in one scene, it is not necessary that the actors are actually able to play—nor are a real cello or piano necessary. In the New York production, they simply sat at imaginary "instruments" and froze in position while a recording played in the background during the scene that followed.

Setting

On board a train from Cambridge to London. The year is 2002.

Act One

(Rory and June are standing on a train platform. Rory looks around nervously.)

RORY: Where is he? The train leaves in ten minutes.

JUNE: I guess he's late.

RORY: He's always bloody late.

JUNE: That's not true.

RORY: Why are you always defending him?

JUNE: I'm not.

RORY: I just wish you could see him as others see him, that's all.

JUNE: I just said that he's not always late.

RORY: Remember the Seattle conference last year? He missed his flight and his lecture on particle radiation had to be postponed until the next day.

JUNE: He had the stomach flu.

RORY: See, that's just what I mean—you're defending him again!

JUNE: What's the matter—are you jealous?

RORY: That's ridiculous.

JUNE: You're jealous.

RORY: How could I be jealous of your husband?

JUNE: Because you're my lover.

RORY: Exactly! He's the one who should be bloody jealous.

JUNE: And yet amazingly, he's not.

RORY: Does he know about us?

JUNE: I don't know. But he wouldn't be jealous if he did.

RORY: Why not?

JUNE: Because he doesn't get jealous.

RORY: That's ridiculous. Everyone bloody gets jealous.

JUNE: You say "bloody" a lot, you know that?

RORY: (with a Cockney accent) That's because I'm bloody working class, ducky. Oh, for Christ's sake, June. What do you mean he doesn't get jealous? Isn't he human?

JUNE: Not quite—no.

RORY: Or maybe it's just too common to get jealous. God knows, George is anything but "common"!

(George enters.)

GEORGE: Hello—sorry I'm late.

JUNE: Hello, darling.

GEORGE: Hello. Hello, Rory.

RORY: Hello, George. We were just talking about last year when you missed your flight to Seattle.

GEORGE: Ah, yes—the stomach virus.

RORY: We were afraid you were going to miss the train. It's almost time for it to leave.

GEORGE: Time is relative, old man.

JUNE: Try telling that to British Rails.

GEORGE: Ten dimensions of space but only one of time.

RORY: What?

GEORGE: M-theory. It's your theory, old man.

RORY: Well, it's not mine, actually; Ed Witten is the one who came up with—

JUNE: We should get on the train.

RORY: Ladies first.

JUNE: If I see a lady around here I'll let you know.

GEORGE: *Touché*, old man.

(They get on the train. The men follow June as she looks for an empty carriage.)

JUNE: This one all right?

GEORGE: Perfectly fine, darling. Whatever you like.

(They sit.)

JUNE: You know, George, there's such a thing as being too agreeable.

GEORGE: Good heavens. (to RORY) Did you ever hear anything like it?

RORY: Women, old man. Best not to try to figure them out. Implicit ignorance—not enough data to go on.

GEORGE: Quite right. (to JUNE) I'm not going to even try to figure you out.

(Pause. There is the sound of a train whistle, and a slight jerk as the train begins to move.)

RORY: I'm glad we were able to get tickets to this play.

JUNE: Yes—it's good to get away from the conference for a while. (Pause. She looks out the window.) Look at the hedges just zipping by.

GEORGE: (to himself) "These hedge-rows, hardly hedge-rows, little lines / Of sportive wood run wild..."

JUNE: What's that, darling?

GEORGE: Wordsworth. "Tintern Abbey."

RORY: George was always rather keen on poetry, even at Cambridge.

JUNE: Ah—even at Cambridge? My goodness.

RORY: I'm not implying that there was no poetry at Cambridge, mind you, but there were...other things.

JUNE: Such as?

RORY: Well...girls—uh, women.

JUNE: So I've heard.

GEORGE: And rugby.

RORY: Ah, yes—rugby.

GEORGE: And rugby women.

RORY: Yes.

JUNE: Women who played rugby?

RORY: Yes...and just women. Those were the days—eh, Georgie?

GEORGE: Yes, indeed—those were the days. (to June) And did you know that Rory was there on full scholarship?

RORY: There's no need to go into that.

JUNE: I'm very impressed.

GEORGE: So you should be. It's nothing to be ashamed of, old man—just because your family couldn't afford to send you—

RORY: Really, George—

JUNE: I agree, Rory—just because George was born into money and you weren't—

RORY: He was also born into the aristocracy, which I decidedly was *not.*

JUNE: If I were you I'd be proud, not ashamed.

RORY: That's because you're not English.

JUNE: I think the English obsession with class is just silly.

GEORGE: That's because you're not English.

JUNE: So *what* if George's family has a title?

RORY: And a castle in Scotland.

JUNE: And a castle in Scotland—so what?

RORY: So what? So everything.

GEORGE: June's right—you should be proud of your accomplishments, old boy. After all, you got a first at Cambridge, whereas I only got a second.

JUNE: I'm surprised you had any time to study at all.

GEORGE: Oh, Rory never studied. He's naturally brilliant.

RORY: It all seems like so long ago.

GEORGE: (to himself) "I grow old...I grow old...I shall wear the bottoms of my trousers rolled."

JUNE: What's that, George?

GEORGE: That is Eliot. As in T.S.

JUNE: Let's not get glum, all right, darling? (to Rory) It's a bad sign when he starts quoting Eliot. Say something to cheer us up, will you?

RORY: All right...right: I've got a joke for you. How many physicists does it take to screw in a light bulb?

GEORGE: (to Rory) I don't know.

RORY: Well, it depends.

GEORGE: On what?

RORY: On whether the light is a particle or a wave.

JUNE: (laughing) Oh, that's a good one.

GEORGE: So how many if it's a particle?

RORY: That's it—that's the joke.

GEORGE: Oh. I see.

(There is an awkward pause.)

RORY: I'm curious to see what this chap is on about. What's his name again, the playwright?

GEORGE: Frayn.

JUNE: Michael Frayn.

RORY: Isn't he the chap that wrote that farce—what's it called?

JUNE: *Noises Off.*

RORY: Right. A bit odd, changing horses in midstream.

JUNE: What do you mean?

RORY: Well, first he's writing frothy little farces, you know, and now he's written this play about physics. That's a far cry from a bunch of birds running around in their knickers. So what's this one about, exactly? What's it called?

JUNE: *Copenhagen.*

RORY: Right.

GEORGE: It's historical.

JUNE: It may be historical, depending on whether he got it right or not.

GEORGE: In any case, the play deals with historical subjects: World War II, physics, the development of the bomb.

RORY: Sounds a bit dodgey to me. I mean, writing about an event that really happened—seems to me you're just setting yourself up for failure.

JUNE: An event that may have happened.

RORY: What do you mean?

GEORGE: A meeting between Werner Heisenberg and Niels Bohr in Copenhagen—

JUNE: Which may or may not have happened.

RORY: That's just silly. It either happened or it didn't.

GEORGE: Uncertainly Principle, old man.

RORY: What?

GEORGE: Heisenberg's Uncertainty Principle—

RORY: I know who bloody Heisenberg was!

JUNE: The play uses that to examine a meeting that may or may not have taken place—

GEORGE: The meeting actually *did* take place. What is uncertain is whether or not Heisenberg deliberately sabotaged the Nazi effort to build an atomic bomb. That's where the connection to Heisenberg's principle comes in—

RORY: Uncertainty only applies to quantum events, which take place on the subatomic level. People aren't atoms.

JUNE: I think it sounds clever.

RORY: Sounds a bit limp-wristed to me.

GEORGE: You don't have to go, old man.

RORY: Oh, no—I'm going all right.

JUNE: What if people *are* like atoms, though? I mean, more than we think?

GEORGE: How so?

JUNE: Well, aren't people as mysterious as the forces in an atom, in their own way?

GEORGE: You mean what if we have more in common with quarks and neutrinos and muons than we think?

JUNE: Yes, something like that.

RORY: I've always felt if I could be any particle I'd be a proton. Positive charge and all that. Also, protons live forever.

GEORGE: As far as we know.

JUNE: What about you, George?

RORY: Oh, George is definitely a neutron.

GEORGE: Am I?

JUNE: Oh, yes. Rory is right.

GEORGE: Why is that?

JUNE: Because you're so neutral about everything.

RORY: What about you, June? What would you be?

JUNE: Maybe I'd be an electron being shared by a couple of big, strong atoms.

RORY: That's a titillating thought.

(George looks at him suspiciously.)

GEORGE: Don't titillate too much, old man.

JUNE: I'm dying for a cup of tea.

RORY: How very British of you.

GEORGE: (to RORY) I sometimes think she's more English than I am.

JUNE: That would be impossible, darling. Anyone else?

RORY: I'll take mine with milk and sugar, please.

JUNE: Right. George?

GEORGE: None for me right now, darling, thank you.

(She leaves. There is a moment of awkward silence between the men.)

GEORGE: I quite enjoyed your lecture on M-theory.

RORY: Oh, thank you. Thanks very much.

GEORGE: I mean it. I found it—stimulating. Not my sort of thing, you know, but—

RORY: What do you mean "not your sort of thing"?

GEORGE: Well, it's not my area of expertise, is it?

RORY: No, I suppose not.

GEORGE: And it's all just speculation, anyway.

RORY: What's that supposed to mean?

GEORGE: Well, I mean, it's all highly speculative, isn't it? You could be right, or you could be whistling through a straw up your ass. No one would know the difference.

RORY: A straw up my ass? What on earth does that mean?

GEORGE: It's just an expression.

RORY: I never heard it before.

GEORGE: You didn't grow up in Kent.

RORY: Neither did you.

(June enters with tea. She hands one to Rory and sits down with hers.)

GEORGE: That was quick.

JUNE: The dining car was empty.

GEORGE: Maybe we're the only ones on the train.

RORY: June, did you ever hear the expression "a straw up your ass"?

GEORGE: That's whistling through a straw up your ass.

JUNE: No. But then I'm a Yank, as you're both so fond of pointing out. We don't have the same finely tuned sense of scatological humor as you Brits.

RORY: Scatologic—

GEORGE: Potty jokes.

RORY: Oh, you think we like potty humor more than you?

JUNE: Definitely. Must be the public schools. All those adolescent boys living in close quarters...

GEORGE: Yes—ours is a deeply wonky society.

RORY: (to George) M-theory is no less reasonable than string theory. In fact—

JUNE: Do you think there's even a remote probability that we could all just enjoy this train ride?

GEORGE: I don't know, darling—probability is more your area.

JUNE: *Touché*, George. What a witty comeback.

RORY: Yes, he should have been a bloody stand-up comic.

JUNE: (looking out the window) This train is really moving.

RORY: Maybe it's the earth that's moving.

GEORGE: That's what June said last night.

JUNE: Maybe it was the train moving last night after all.

GEORGE: No, that was actually the Big Bang.

RORY: Very funny, both of you. (to George) So what exactly is your problem with M-theory?

GEORGE: First of all, why do you call it M-theory? That's so irritating. Why not just call it Membrane theory?

RORY: You'd have to ask Ed Witten. He's the one who came up with it.

JUNE: A lot of people say he's the greatest physicist since Einstein.

GEORGE: But even Einstein couldn't solve the problem of the singularity at the Big Bang—

JUNE: The way the laws of physics break down at that moment.

RORY: I wonder what it's like to be that smart.

JUNE: Oh, come on, Rory—you got a scholarship to Cambridge, for God's sake! You're smarter than either George or me.

GEORGE: I beg your pardon.

RORY: Still, to have discovered M-theory...

JUNE: If you could be any physicist from history, who would you be?

GEORGE: Dead or alive?

JUNE: Either one.

RORY: Like Schrödinger's cat—both dead and alive.

GEORGE: Isaac Newton.

JUNE: That's boring.

GEORGE: He was like Columbus! Nobody knew anything before he came along.

JUNE: Except Galileo.

GEORGE: He wasn't the mathematician Newton was.

RORY: Newton was also an arrogant ass.

GEORGE: He was the Father of Physics. He had a right to be arrogant.

RORY: Did he tell you that himself?

GEORGE: Maybe. What if he did?

JUNE: What about you, Rory? Who would you be?

RORY: Max Planck. I've always wanted to be Max Planck. Not Heisenberg, with his messy uncertainty, but Planck...to discover something like the Planck Constant. (Dreamily) The Planck Constant...the birth of Quantum Physics.

GEORGE: What about you, darling?

JUNE: That's easy. Marie Curie.

GEORGE: Because she was a woman?

JUNE: No, because she was heroic. They both were, Marie and Pierre, but she carried on after his death. Even though she knew she was being slowly poisoned by uranium, it didn't stop her from doing her work.

RORY: There is something noble about her, and something magical about her relationship with Pierre.

JUNE: Hey—I overheard someone at the conference say that the M in M-theory actually stands for "magic."

GEORGE: Or madness.

JUNE: Mystery.

GEORGE: Matrix.

JUNE: Mother.

GEORGE: The Mother of All Theories.

RORY: Call it whatever you want. What's your problem with it? My equations were all solid.

GEORGE: As far as we know.

RORY: What's that supposed to mean?

GEORGE: You still haven't solved the problem of the Big Bang – what exactly banged, and why. But your equations are very elegant. I think we all agree you're a superior mathematician.

JUNE: Well, he is a particle physicist—they *are* the math whizzes, after all.

GEORGE: That's true. We cosmologists can't hold a candle to your—

RORY: Oh, come off it, George! You're a string theorist, for god's sake! We both know the math in string theory is devilish wicked.

GEORGE: Oh, but M-theory is so trendy just now; it's the Next Big Thing. What a charming concept: all matter sitting on these subatomic membranes floating around like giant bedsheets. And we're sort of like fleas hitching a ride, clinging on for dear life.

RORY: Well, string theory *is* getting a bit tired, isn't it? I mean, you string theorists are all—forgive me—rather tied up in contradicting theories.

JUNE: (singing) Fermions, bosons, all tied up in strings—

RORY: (singing) These are a few of my favorite things.

JUNE: (singing) Neutrinos in spandex and quarks in white dresses—

RORY: (singing) Hadrons colliding and making big messes—

JUNE: (singing) Leptons and mesons in tight little rings—

BOTH: (singing) These are a few of my favorite things.

GEORGE: Very funny, both of you.

RORY: I don't see that M-theory is any more speculative than string theory—for god's sake, you string theorists can't even agree with each other! How many competing theories are there now—six, seven?

GEORGE: Five.

RORY: Five!

GEORGE: Competing theories help build knowledge.

RORY: But *five* competing theories? And they say three's a crowd.

JUNE: Oh, I don't know...it takes three quarks to make a proton or neutron—

GEORGE: "Three quarks for Muster Mark."

(Rory stares at him.)

GEORGE: James Joyce—*Finnegans Wake.*

JUNE: And you need an electron, proton and neutron to make up an atom—

GEORGE: And three to make a triangle. Quite a stable geometric shape, a triangle...

(There is an uncomfortable pause.)

RORY: I think I'll go out for a stroll.

(He leaves.)

JUNE: What's wrong with Rory? Have you been baiting him again?

GEORGE: I just mentioned that M-theory was in the early stages yet.

JUNE: Oh George, you know how sensitive he is.

GEORGE: Why does Rory get to be the "sensitive" one? Why don't I get to be sensitive?

JUNE: Because you're not. You're a rock, hewn in granite—a big, solid boulder. Level-headed, sensible—but not sensitive.

GEORGE: It's not fair.

JUNE: Well, darling, life isn't fair. You should have married a nice girl, but you're stuck with me.

(Pause.)

GEORGE: June?

JUNE: Yes, George?

GEORGE: Do you...do you remember that one rainy weekend in New York, when we were living in Hell's Kitchen—when we seemed to be a universe unto ourselves? The world began and ended with the four walls of that apartment.

JUNE: We never left those three rooms all weekend.

GEORGE: And life never felt so complete, so—full as it did that weekend.

JUNE: I remember.

GEORGE: It was as though we had created our very own dimension in space-time...like we had found something fundamental, and were part of a great universal experience. It was like...physics.

JUNE: You read me poetry.

GEORGE: Yeats, Coleridge, Rilke...and I never felt the need for other people. Did you feel that too? Or was it just me?

JUNE: I remember.

GEORGE: I've actually tried to forget, but I remember.

JUNE: Why would you try to forget?

(Rory enters.)

RORY: Am I interrupting something?

GEORGE: No.

RORY: Did you know that the penis of a humpback whale is twelve feet long?

GEORGE: Good God, Rory.

JUNE: What made you think of that?

RORY: I just saw this big hill out the window, and it reminded me of a humpback whale. And I remembered I had seen this special on the Science Channel about whales...(to George) so if you laid us end to end, we still wouldn't be as long as the penis of a humpback whale.

GEORGE: All right, Rory—I get it.

RORY: You're cranky. (to June) Why is he so cranky?

JUNE: He missed his nap time.

RORY: Oh, we missed our nappies, did we?

GEORGE: Don't push your luck, old man.

RORY: My luck is the last thing I'd be likely to push.

JUNE: Oh, speaking of nappies, I understand your sister is having a baby!

GEORGE: I always said that woman was a breeder.

JUNE: George! (to Rory) I'm so glad for her. I'm sure she'll be a wonderful mother.

RORY: Yes, no doubt…

JUNE: Do you…

RORY: What?

JUNE: Do you ever think about having children?

RORY: I'd have to get married first.

GEORGE: Take my advice, old boy—don't.

RORY: I was in the library one day last fall, and there was this tiny girl—I don't know how old she was—three, four? – a perfect human being in miniature. She was struggling with an enormous pink backpack that was almost as big as she was. Her mother was across the room, putting some books back on the shelves. I stopped to help her, and as I caught sight of these tiny, perfect hands I was suddenly overcome by longing—an actual physical ache, the kind you feel when you're in love. I felt light-headed with this untidy jumble of emotions—happy and sad all at once. It was like a fireworks of chemicals in my brain had been triggered by some ancient, instinctive receptors. I wanted to throw a protective web around her and keep her from all things bad and harmful in the world. I don't remember her face – I suppose it was pretty; most children that age are pretty—but I'll never forget those incredible tiny hands. I knew then what mothers feel…it's fierce and powerful and frightening.

GEORGE: So then what happened?

RORY: I helped her on with the backpack, her mother smiled at me, and off they went.

GEORGE: That's it?

RORY: What do you mean?

GEORGE: You didn't kidnap her or anything?

RORY: Why would I do that?

GEORGE: That's a disappointing ending.

RORY: (to June) Do you think it's disappointing?

JUNE: (to George) What *is* it with you?

GEORGE: I'm a man! I want simple, action-filled stories.

JUNE: I have more testosterone than the two of you combined.

RORY: She has a point, old man.

GEORGE: That's ridiculous!

RORY: She is a rock climber, old man.

GEORGE: Your rock climbing is rather ironic, don't you think, considering your breakthrough theory on why gravity is so weak.

JUNE: It may be the weakest of the four forces, but it sure doesn't feel that way when you're scaling El Capitan.

GEORGE: What's the fun in it?

JUNE: I feel like I'm experiencing and defying gravity all at the same time.

GEORGE: But it looks so—tedious.

JUNE: It's not about the progress—it's about the process.

RORY: What do you like about it?

JUNE: What do you like about playing music?

RORY: It's...a conversation. Between me and the composer – and between me and the other musicians.

JUNE: Rock climbing is a conversation with the rock.

GEORGE: Really, darling, that is a bit—

JUNE: There's an abstract beauty about the way the crags join together…it's like the beauty of mathematics.

RORY: But it's so dangerous.

JUNE: And it forces me to be totally in the moment. Time doesn't exist—there's only now. No past, no future. I don't ruminate or plan—I can't afford to.

RORY: And you like that?

JUNE: I can leave my life behind and just be a part of the rocks.

RORY: Is your life so terrible that you want to leave it behind?

JUNE: No, but it's confining and confusing and petty and…human. I like being part of something larger, to just be a speck of flesh and bones crawling up this huge mountainside.

GEORGE: That actually sounds disturbing.

JUNE: I feel like I want to—to know the rock itself. Each rock invites me to solve the puzzle of how to climb it, only I'm solving the puzzle with my whole body, not just my mind. On a really good day, I blend into the mountain—I become the rocks.

GEORGE: This is getting a bit too strange for me.

JUNE: But we're all made of the same material as the rocks, more or less.

RORY: Yes, that's what M-theory is on about, you see! Everything—music, flowers, sunsets, the mystery of love, the whole range of bloody "meaningful" clichés of beauty and truth—it's all getting at that, the center of things, the fact that we're all part of this—this—

JUNE: Membrane.

GEORGE: Waving around in subatomic space.

RORY: Yes. I read your paper, June—I thought it was quite brilliant, the idea that gravity is leaking into our universe from a parallel one. It dovetails so well with M-theory—

GEORGE: So if you M-theorists are right, then there are parallel universes tucked away in between ours—

JUNE: Even Emmanuel Kant proposed the existence of parallel planes, which he called "separate worlds."

RORY: There may be a universe in which June is married to me, for example, instead of you.

GEORGE: In your dreams!

JUNE: And there may be a universe somewhere, floating on a membrane just next to ours, in which the Twin Towers never fell, and David and I are seated at Windows on the World having breakfast looking out at the bluest September sky I can remember.

(George looks at her, obviously stung.)

JUNE: I'm sorry. That was wrong of me, to bring that up. I'm sorry.

(George is still silent.)

JUNE: George? I'm sorry. George?

GEORGE: Let's just *forget* it, all right? (Pause) I think I'll go out for a little air.

(George leaves. There is a pause.)

RORY: I think he knows.

JUNE: What?

RORY: About us.

JUNE: Don't be paranoid.

RORY: I'm not. I really think he knows.

JUNE: Why do you say that?

RORY: He's being so passive/aggressive.

JUNE: That's just George.

RORY: You saw the way he was coming at me about M-theory.

JUNE: He was just being playful.

RORY: He was making fun of me.

JUNE: He respects you.

RORY: Really? You think so?

JUNE: Do you think I could sleep with you if my husband didn't respect you?

(George enters.)

GEORGE: It was wrong of me to walk out like that. I—I should have been bigger than that. (to June) I'm sorry.

JUNE: It's all right, George. I'm sorry I'm brought it up.

(Pause.)

JUNE: Christ, I need a cigarette.

RORY: I thought you quit.

JUNE: I did. I'll have to go borrow one from someone.

(She leaves. Pause.)

RORY: She didn't mean anything by it, you know.

GEORGE: She's still struggling with David's death. I know that.

RORY: I can't imagine what it was like for the two of you to lose your only child like that—

GEORGE: Why don't we just drop it, all right?

RORY: Why aren't you struggling with it like she is?

GEORGE: It was God's will.

RORY: If you don't mind my asking, how do you reconcile your faith with—

GEORGE: With what?

RORY: With being a scientist?

GEORGE: I don't see a conflict. Even Einstein said, "God does not play dice."

RORY: He was speaking metaphorically.

GEORGE: How do you know?

RORY: Einstein wasn't a Catholic.

GEORGE: No, but he was a Jew. Same thing—both are quite keen on guilt.

RORY: Why do you need religion when you have science?

GEORGE: I don't *need* religion; I just happen to believe.

RORY: What do you find so compelling about Catholicism?

GEORGE: I like the incense.

RORY: Seriously.

GEORGE: You sound envious.

RORY: Maybe I am.

GEORGE: I like the Holy Trinity.

RORY: The Father, the Son, and the—what exactly is the Holy Ghost?

GEORGE: I once asked my mother that and she said it was kind of like Father Christmas.

RORY: You mean the Holy Ghost brings presents to good little Catholics?

GEORGE: No, more like it was the spirit of Christmas was not about the presents, but the spirit of giving and all that.

RORY: That doesn't mean much to a five year old who really wants a fire truck.

GEORGE: Speaking of which, there's a question I've been wanting to ask you.

RORY: Go ahead—shoot.

GEORGE: That's an unfortunate choice of words.

RORY: Why? What's the question?

GEORGE: I wanted to ask you if you've enjoyed sleeping with my wife.

RORY: Well, I—

GEORGE: That's a stupid question—of *course* you've been enjoying it; otherwise, you wouldn't be doing it. At least I assume you're enjoying it—if not, that rather puts your motives in question, doesn't it?

RORY: Look, George—

GEORGE: I mean, what reason could you possibly have for sleeping with a woman if you're not enjoying it? It would be rather insulting to her, don't you think? Unless you're doing it to spite me, but I don't really think that's the reason. That doesn't seem like you at all, does it?

RORY: Look, can I just—

GEORGE: We've never been enemies, have we? In fact, I once thought we were friends. You don't have some hidden axe to grind with me, do you? That redhead at the rugby party back in our Cambridge days, perhaps? She really did fancy me over you, you know.

RORY: For God's sake, George—

GEORGE: I can only assume you're enjoying sleeping with June because I quite enjoy it myself. Or rather, I used to enjoy it—not so much of her to go around anymore. She is rather like an electron being shared by two atoms, isn't she? So that makes the three of us an odd kind of molecule—

RORY: Could I just say something—

GEORGE: I suppose the real question is, are you planning on continuing it, and if so, for how long do you plan to continue?

RORY: We haven't planned anything—we didn't *plan* to sleep together in the first place—

GEORGE: Just chemistry, eh?

RORY: Something like that.

GEORGE: So there you were, a happy little hydrogen atom, just floating along minding your own business, feeling a bit thirsty perhaps, thinking of bonding with some juicy little oxygen atom, when along comes this absolutely knockout electron cloud, and you can't resist—wham! The two of you make a big, fat water molecule. Suddenly, there it is: ladies and gentlemen, chemistry in action!

RORY: I wish you would stop being so arch about it.

GEORGE: Oh, you do? *You* wish *I* would stop being arch? *Really*?

RORY: Look, we—we were about to break it off.

GEORGE: Is that supposed to be a comfort to me?

RORY: I don't know; we both feel terribly bad about it.

GEORGE: Well, isn't that splendid? That's just terrific.

RORY: June said you weren't the jealous type.

GEORGE: Did she? And just exactly what type I am supposed to be?

RORY: I don't know.

GEORGE: She doesn't love you, you know.

RORY: I'm not sure if I care.

GEORGE: Bullocks!

RORY: What is "love," anyway? Eye of the beholder, and all that.

GEORGE: So you were about to break it off? What's the matter, old boy? Finding it a bit hard to perform under pressure? You know what they say: what goes up must come down.

(Sir Isaac Newton appears. The lights go up on him and down on Rory.)

NEWTON: I never said that, you know.

GEORGE: Yes, I know. I didn't mean to imply—(suddenly realizes who he's talking to)

NEWTON: In any event, it appears now that there is no up and down—except in your electrons. Nasty little things. Up, down, spin, half spin—it all renders me quite dizzy.

GEORGE: Excuse me, but I was in the middle of a very serious—

NEWTON: Shall I tell you the worst thing about being the greatest scientist who ever lived?

(Pause. George stares at him.)

NEWTON: Well?

GEORGE: Oh, yes—yes, please do.

NEWTON: It is not being fully understood. Most people think of me and they think, "Oh, yes, Newton—isn't he the chap who discovered gravity? Saw an apple falling and all that – seems pretty simple to me, it's a wonder it hadn't been done before. Downright obvious, in fact; why, I could have come up with that if I were idle enough to sit around watching apples fall all day long. What's the big deal about this chap

Newton, anyway?" Can you *possibly* imagine how *infuriating* that is to me?

GEORGE: No...I mean, yes.

NEWTON: Most people completely fail to grasp the significance of my contribution—my genius, if you will.

GEORGE: Yes, yes—quite.

NEWTON: After all, I invented physics!

GEORGE: Well, there was Galileo—

NEWTON: Italian poufter. Played around with dropping balls off towers. And it turns out that story was a fake—it never actually happened. I am *twice* the mathematician he was.

GEORGE: He invented the telescope—

NEWTON: Do you know how he did that? It was a child's toy, and he adapted it to look at the stars. *I'm* the one who added the mirror and made it really useful. And his "big theory" of the earth travelling around the sun had already been discovered by Copernicus, you know.

GEORGE: Yes, I know. Did you really come up with gravity from watching an apple fall?

NEWTON: Actually, it was a peach.

GEORGE: Really?

NEWTON: Yes, but I thought an apple sounded more Biblical. Oh, I have a joke for you. How many seventeenth century physicists does it take to screw in a light bulb?

GEORGE: How many?

NEWTON: What's a light bulb? (Pause) Get it?

GEORGE: Yes! Very good. (Pause.) Can I ask you a question?

NEWTON: You may. After all, I have all eternity to kill.

GEORGE: What about...

NEWTON: What? Speak up, dear boy.

GEORGE: What about sexual attraction? Have you solved that mystery yet?

NEWTON: Physics. It's all physics. Your atoms responding to her atoms. Or, if you want to be crude, chemistry. You're a scientist, old boy—you understand the significance of my work, don't you?

GEORGE: Yes—oh, yes! I do.

NEWTON: Of course you've scrapped my theories now...

GEORGE: Oh, no—we've built upon your theories!

NEWTON: Don't try to comfort me. First there was that wretched German with the fuzzy hair and mustache, and now all of this uncertainty business...where will it lead us? And string theory! Goodness...where is the Divine Hand in that, I ask you?

GEORGE: I'm convinced it's there somewhere.

NEWTON: Where do you see God in modern physics?

GEORGE: Everywhere. I don't see why science and religion should be always at odds.

NEWTON: It seems to me you're moving farther apart all the time. Ptolemy put God and his greatest creation, man, at the center of things. But Copernicus—

GEORGE: Copernicus was a Catholic cleric.

NEWTON: Everyone with an education was a cleric in those days. But that's also the same church that persecuted Galileo.

GEORGE: I'm a Catholic.

NEWTON: So you go in for all that Mother Mary fetish, do you?

GEORGE: I wouldn't exactly call it a fetish. You yourself are—

NEWTON: Church of England, dear boy. That grand religion born of the lust and ambition of a ruthless, wife murdering monarch. So you see God's hand in—all of this?

GEORGE: Just look at the beauty of the design—the divine symmetry of it all! How could that be anything other than the mind of God?

NEWTON: Tyger Tyger, burning bright, / In the forests of the night; / What immortal hand or eye, / Could frame thy fearful symmetry?

GEORGE: William Blake wrote that two hundred years after your death! How did you—?

NEWTON: I've managed to keep up. I thought you would appreciate it, being a connoisseur of poetry…and he *was* English, after all. Sadly misguided, of course, but what do you expect from a Christian mystic?

GEORGE: Ah, yes—the "God within"!

NEWTON: What good is a god like that? It's like having indigestion.

GEORGE: Symmetry is a key to our understanding of physics today.

NEWTON: Ah, yes: "a property of a physical system that does not change when a system is transformed in some manner."

GEORGE: Do you know what supersymmetry is?

NEWTON: Enlighten me.

GEORGE: It's the theory that every particle has an as-yet undiscovered superpartner.

NEWTON: It's just a theory.

GEORGE: For now.

NEWTON: Hmmm…it looks to me as though you have your supersymmetry partner.

GEORGE: You mean Rory.

NEWTON: I can see the horns on your head. (Chuckles) Metaphorically, of course. Do you still use that expression—cuckold?

GEORGE: Look, if you don't mind—

NEWTON: I can take a hint! (Sighs) And I thought you would be delighted to see me.

GEORGE: On another occasion, perhaps, but your timing—

NEWTON: Time never was my strong point. Just ask that horrible Einstein. He has time all figured out, doesn't he—thinks he's such a big "hot shot"—what a great genius. "God does not play dice with the universe"—ha! Not only does he play dice; I'll tell you a secret: he has a major gambling problem! Don't tell anyone I told you.

(He leaves. Lights go back up on George and Rory just as June enters.)

JUNE: So, what have you two boys been up to while I was gone?

(Silence. Rory and George avoid looking at each other.)

JUNE: What have you been talking about?

GEORGE: M-theory.

JUNE: I hope you've been nice to Rory. (Pause) Well—have you?

RORY: Yes, he's been very nice.

GEORGE: Liar. I was a real shit.

RORY: Okay—you were a real shit.

(He exits.)

GEORGE: Or maybe I've been a little bit of both, depending upon who the observer is. In the quantum world, I can't really be said to be anything unless someone is observing me.

(A blackboard appears to one side of the stage. George steps over to the blackboard as the lights go up on the blackboard and down on Rory and June. On the blackboard is a diagram of the experiment of Schrödinger's Cat, which George refers to during the following lecture he delivers to an imaginary class of students.)

In the mid 1920's, in order to demonstrate the weirdness of the quantum world, Erwin Schrödinger came up with the following thought experiment. He asked us to imagine a cat in a box. Inside the box with the cat was a quantum event of some kind—in other words, a subatomic process—say, a radioactive isotope which might or might not decay in a given time, emitting a beta or alpha particle. In the theoretical world of the experiment, the decay of the isotope would trigger the release of a vial of poison inside the box, and the cat would die. However, if the particle was not released and the decay did not take place, the cat would live. As we know, atomic decay is a matter of probabilities, so even though the isotope has roughly a fifty percent chance of decaying, it is by no means certain—any more than drawing four aces in a hand of poker is certain. So far, so good.

But here's where the weirdness of the quantum world comes in. According to Schrödinger, the cat could not be said to be *in either state* until an observer actually *opened* the box and looked inside. The cat's condition would then "collapse" into an observable state—dead or alive—in the same way an electron's wave function can be said to "collapse" into a particle the second you shine a photon on it.

This experiment was purely theoretical—no cats were harmed, or even used. And the subatomic event could be anything where quantum mechanics would apply—it didn't necessarily have to be the decay of a radioactive isotope. So if you take quantum mechanics at its purest level, the cat cannot be said to be dead or alive *until it is observed*—until that moment, it exists in an indeterminate state, neither dead *nor* alive.

Schrödinger meant it to sound as absurd then as it does to us now, but the implications of this are profound—what it says essentially is that we co-conspire with our environment to create reality. And since then, we have discovered that all matter does in fact have a wave function, so the absurdity of it is not as great as Schrödinger imagined.

(Newton enters.)

NEWTON: *Esse es percipi.*

GEORGE: "To be is to be perceived." Did you say that?

NEWTON: Good heavens no! George Berkeley said it.

GEORGE: The philosopher—your contemporary?

NEWTON: The same. Fellow actually believed that everything that exists is either a mind or depends upon a mind for its existence. Can you imagine that? He thought everything was an illusion, and that matter doesn't exist. He believed that ordinary physical objects are composed solely of ideas, which are inherently mental.

GEORGE: That's not so far off from what we've learned in quantum theory, you know.

NEWTON: So then the question becomes, did your friend sleep with your wife or didn't he? If you didn't see it, perhaps it never happened.

GEORGE: But now in my mind I see it happening over and over again.

NEWTON: Who's to say which is more real—your nightmares or their actions in or out of bed?

GEORGE: In cosmology, we've learned that the farther away galaxies are from one another, the faster they're moving apart. I feel like that's happening with June and me—and I don't know how to bring her back.

NEWTON: I wish I could help you, dear boy, but I was never very good with people.

(The lights go down on Newton and blackboard and up on June. George goes over to her. Rory is frozen in semi-darkness, as though he is not there.)

GEORGE: Why, June? Why did you do it?

JUNE: Because I'm a very bad person.

GEORGE: What is it about Rory you find so attractive?

JUNE: He's...smutty.

GEORGE: What?

JUNE: Sex with Rory is...dirty. It's low and sensual and...erotic.

GEORGE: And with me? Are you saying it's not erotic with me?

JUNE: No, it's not that—

GEORGE: What, then?

JUNE: Look at yourself, George. You're—you're so *virtuous.*

GEORGE: What do you mean?

JUNE: You're so *good*—you're a good Catholic, a good husband, a good scientist—don't you ever get tired of being so damn good all the time?

GEORGE: For pity's sake, June—

JUNE: Don't you ever feel like letting your hair down and...I don't know—stealing a hotel towel or something?

GEORGE: It's never occurred to me.

JUNE: Exactly! The rest of us are fighting temptation every day, but it doesn't even *occur* to you! You even wear your seat belt if you're just going to the store for milk!

GEORGE: It's the law.

JUNE: But I bet you'd wear it even if it weren't the law!

GEORGE: Probably.

JUNE: Why?

GEORGE: It's safer.

JUNE: Exactly. You're...*safe*, George.

GEORGE: That's why you had sex with Rory? Because I'm *safe*?

JUNE: Something like that. I think I'm not worthy of you. So I go and sleep with Rory because he's no better than I am.

GEORGE: Good lord. There's no defense against that.

JUNE: I know. And I hate myself.

GEORGE: I'm not perfect, you know. I'm often late...

JUNE: That's not dangerous, that's predictable! Oh, god—listen to me! I'm cheating on you and the best reason I can come up with is that you're too *good*! That's pathetic. Jesus Christ. Jesus Fucking Christ.

GEORGE: June, I wish you wouldn't—

JUNE: I know—I know—you wish I wouldn't take the Lord's name in vain. Well, I wish I wouldn't do a lot of things but somehow I seem to keep doing them anyway! I just...I just can't—I can't leave you.

GEORGE: So maybe you can force me to leave you. Is that it?

JUNE: Oh, for Christ's sake, George, will you stop *analyzing* everything? It's exhausting.

GEORGE: But why, June? Why did you do it?

JUNE: Because I shouldn't even be here! I should have died with David—I should have died with our son!

GEORGE: Don't you *say* that—don't *ever* say that!

JUNE: You know it's true! I was supposed to be there too...

GEORGE: Stop it, June! For God's sake—

JUNE: I cannot accept what happened. I can't wrap my mind around *why* it had to happen.

GEORGE: There is no Why. It was God's will, that's all.

JUNE: How can you believe that if He exists, He cares anything about us? He left us here—abandoned like a box of puppies left on the side of the road!

GEORGE: David was my son too! Do you think I don't grieve for him every minute of every day?

JUNE: I know you do—but *you* don't blame yourself for his death!

GEORGE: How can I, when I believe it was God's will—

JUNE: You know, your damn faith is like a slap in the face! (Pause.) I'm sorry, George, but I don't think I can talk to you about this.

(Marie Curie appears, and the lights go up on her and down on George.)

MARIE: Perhaps you can talk to me?

JUNE: Madame—Marie Curie?

MARIE: I too lost someone dear to me. An accident, I know – Pierre fell under the hooves of a panicked horse.

JUNE: Yes, I know. We've all read about your tragedy.

MARIE: So I know something of your grief, because I felt it too. Why was the road wet, why did the horse panic, why did Pierre have to slip and fall at that moment?

JUNE: In string theory, the fabric of space-time *can* repair itself! So why couldn't it curve around itself and protect him from what fell out of the sky that day?

MARIE: Over and over you play it in your mind. Like a sickness.

JUNE: I just can't seem to move past it.

MARIE: Well, these things take time...even in your new world of relativity and all that. Grief has its season, no?

JUNE: How did you get past it?

MARIE: I think I had to learn to make it a part of me—the sorrow. I finally stopped fighting it and embraced it—like it was my child.

JUNE: Didn't you *miss* him?

MARIE: Yes.

JUNE: I'm so *angry.*

MARIE: Yes, I had that too, for a long time.

JUNE: I feel like I don't deserve to be happy, ever again. I...have other feelings, too, ones I'm ashamed of.

MARIE: That's sad, eh, to be ashamed of your feelings. They are just feelings.

JUNE: Two contradictory feelings at the same time.

MARIE: But isn't that just like this Uncertainty Principle you are all talking about?

JUNE: I haven't told anyone about it because it—it confuses me.

MARIE: Well, so it confuses you. Again, just a feeling. But feelings can poison you as surely as radioactive isotopes poisoned me.

JUNE: It was in the time period...afterwards. We were stunned, we were all grieving, but there was a kind of...transformation. Instead of competing with each other, we wanted to take care of each other. Instead of fighting over every little thing, claiming our territory on buses, on subways, on elevators, we gave up our seats; we were suddenly gracious and graceful and giving. I felt so proud to be a New Yorker. Overnight, like a miracle, we were connected. This city of jostlers and hustlers became an island that felt...sacred. The horror was followed by this sweet, magical time where we all felt as though we had been touched by grace. And everyone loved us. The Germans, the Italians—even the French—they all suffered with us. The rest of the country who had always hated and resented us, suddenly felt sorry for us. Even Texas loved us—at least for a while. Ever since then I've longed for that sense of connectedness.

MARIE: Perhaps it is that longing that attracts people to organized religion. They want to be part of something more important—something sacred. There's something in us that reaches after that, the sense of the sacred, no?

JUNE: I had just lost David, and I missed him so much. But then there was this...connection. I didn't know what to feel.

MARIE: Your husband—he has the faith, eh? What about you?

JUNE: No, I don't believe. Especially not after...what happened.

MARIE: Not right for a scientist to believe in God, eh?

JUNE: No, I want to believe. As a kid I remember going to church and waiting for the Mystical Religious Experience. But I just can't believe in anything I don't *know*.

MARIE: Do you really know that the nucleus of a radioactive atom will decay?

JUNE: Well, not exactly, but we know there's a probability—

MARIE: Ah—a probability. Why can there not be a probability of God existing?

JUNE: I just don't think he does.

MARIE: So it's a belief.

JUNE: Yes.

MARIE: Just like faith.

(Pause.)

JUNE: That day...I was supposed to meet David at Windows on the World. His company was throwing a big breakfast, and he knew I'd always wanted to see the city from up there. No fear of heights for me—with all the rock faces I've climbed: Devil's Peak, Breakneck Ridge, Dead Man's Hollow...

I was hurrying to get there, but I lingered just a few moments over a second cup of coffee. I'd forgotten to fill my Metro Card, and there was a line and I couldn't find my credit card; I finally found it but I had to wait for the next train because the first one was too full. I remember standing on the platform, my forehead burning with rage and impatience because I was going to be late, and I didn't want to keep David waiting. When I finally got a train it was almost nine, and I squeezed in with everyone else, holding my breath between stops. It seemed to take forever and the train was diverted to another station.

The minute I got off I knew something was wrong. There were people coming down the stairs and they looked frightened and upset and when I got upstairs I saw this woman sitting on the top step crying. Everyone else was just standing there looking up…and then I saw what they were looking at. The second plane had just hit Tower Two, and smoke was pouring out of Tower One, in giant plumes, and it was only when the second plane hit that people were beginning to realize that this was no accident, that someone meant to do this.

And I also knew I should have been up there with him when it came crashing down. Those faces on the posters that turned up over the next few weeks—hundreds of them, thousands—plastered all over bus stops, trees, lamp posts, park benches. Faces, in black and white and color, laughing, smiling into the camera—all ages and races, but always the words on the posters were the same: Missing. Please Help. (Pause.) *I* should have been one of those faces.

MARIE: Was he ever on a poster?

JUNE: No. I knew he was gone the minute I saw the second tower fall. It was almost like…

MARIE: Like what?

JUNE: There is this thing called quantum entanglement. Einstein called it "spooky action at a distance"—have you heard of it?

MARIE: No.

JUNE: It's a given of Einstein's theory that nothing can travel faster than the speed of light—right?

MARIE: Yes.

JUNE: But we've found something that can. Two particles created at the same time can communicate faster than the speed of light.

MARIE: How can that be?

JUNE: We don't know, but it's been proven in countless experiments. If you shine a photon beam on an electron, its wave function will collapse and it will become a particle with a given spin.

MARIE: I see.

JUNE: And no two related electrons can have the same spin.

MARIE: Ah. Rather than compete with each other's spin, they must complement each other.

JUNE: Right. Well, the moment the first electron "declares" its spin, the second electron will immediately assume the opposite spin—thus, they are communicating instantaneously, faster than the speed of light.

MARIE: *C'est incroyable*—that's incredible.

JUNE: I felt like one of those electrons that morning—I just knew, absolutely knew, that he was gone—instantly.

MARIE: Spooky action at a distance.

JUNE: Yes.

MARIE: This world, it's a dance, eh? All we can do is keep dancing…and live with the uncertainty. The nucleus of a radium atom may decay, but there is always the outside chance it may not—and that is a possibility we have to cling to.

JUNE: I just—

MARIE: What?

JUNE: I have these longings…for certainty, for knowability.

MARIE: That is why you became a scientist, no?

JUNE: I suppose. I don't even know what I long for…

(The lights go down on Marie Curie and up on George and Rory.)

GEORGE: Something…

RORY: Something to do with time—

GEORGE: Space—

JUNE: —and time.

RORY: More space—

GEORGE: More time—

JUNE: A particular kind of time—

GEORGE: What *is* time, after all?

JUNE: Is it just an illusion?

RORY: Stretching out in front of me—

JUNE: —endless as the oceans—

GEORGE: Just a prison, time and space—

JUNE: Infinity is out there somewhere—

RORY: Knowledge is out there somewhere—

GEORGE: Faith—

JUNE: Hope—

RORY: Charity—

ALL: Love.

GEORGE: What happens to love?

JUNE: I want—

GEORGE: I want—

RORY: I want everlasting love—

GEORGE: Eternal life—

JUNE: Youth and beauty—

RORY: Truth and goodness—

GEORGE: A god I can depend on—

JUNE: A really good FM station—

RORY: Truth in advertising—

JUNE: Another Sting album—

RORY: Cheap, affordable housing for everyone—

GEORGE: A really effective deodorant.

JUNE: I want multiple orgasms—

GEORGE: Coffee that tastes as good as it smells—

JUNE: A diet pill that really works—

RORY: —socialized medicine—

GEORGE: —the return of liberalism—

JUNE: Cable guys who show up when they say they will.

GEORGE: The outlawing of junk mail—

RORY: —the outlawing of the Royal Family—

GEORGE: —of cellphones—

JUNE: —of NASCAR. I want a horse who knows the way home—

GEORGE: —and a woods to ride her in—

RORY: I want politicians who don't lie to you—

JUNE: Salesmen who don't lie to you—

GEORGE: Wives who don't lie to you.

JUNE: I want a really good back rub—

GEORGE: —a really good hot bath—

RORY: —a really good blow job.

JUNE: I want a vaccine for the common cold.

GEORGE: A vaccine for stupidity—

RORY: A vaccine for conservatives.

JUNE: A chicken in every pot—

RORY: Pot for everyone—

GEORGE: Chicken pot pie.

JUNE: I want another winning season for the New York Mets.

RORY: I want more time—

JUNE: Quality time.

RORY: What the bloody hell is "quality time"?

GEORGE: What happens to love between two people?

RORY: Where does it go when we stop believing?

JUNE: Where does it go?

GEORGE: Why do we have this nagging sense—

RORY: —that there's *something else...*

JUNE: Tantalizing intimations of—

GEORGE: Glory.

RORY: Why do people need to believe?

GEORGE: Is belief itself a cruel joke?

JUNE: I believe in the scientific method, the transformative power of love, the late Beethoven quartets—

GEORGE: —the creativity of the human spirit, the destructive and healing power of Nature—

RORY: (overlapping) I believe in rock and roll, in desire between a man and a woman—

JUNE: Or between two men or two women—

GEORGE: —or between two muskrats—

RORY: —whatever floats your boat—

GEORGE: (overlapping) I believe in the importance of great books, the collective unconscious, the healing power of music and art—

RORY: —the healing power of really good wine –

JUNE: —the healing power of love—

RORY: —the healing power of really big tits.

JUNE: I want—

RORY: I believe—

GEORGE: I want to believe.

RORY: I believe I have wants.

JUNE: It's a dance.

RORY: All we can do is keep dancing.

JUNE: Keep dancing—

GEORGE: Until the music stops—

RORY: Or the train stops—

JUNE: Keep dancing—

RORY: Cheek to cheek—

JUNE: Keep dancing—

GEORGE: In the dark—

ALL: Keep dancing.

Act Two

(Rory and June are standing on a train platform.)

RORY: Where is he? The train is leaving in ten minutes.

JUNE: I guess he's late.

RORY: He's always bloody late.

JUNE: That's not true.

RORY: Why are you always defending him?

JUNE: I just said he's not always late.

RORY: Remember the Seattle conference last year? He missed his flight and his lecture had to be postponed until the next day.

JUNE: So I heard. But you said the lecture went well.

RORY: See, that's just what I mean—you're defending him again! You'd think you were married to him, for Christ's sake.

JUNE: Don't be silly, darling—I'm married to you.

RORY: Oh, yes...why did you marry me again?

JUNE: Because of your enormous pe—

RORY: No, I'm serious.

JUNE: What kind of question is that?

RORY: I always had the feeling you preferred George.

JUNE: Don't be silly.

RORY: No, I mean it. Did you prefer George?

(George enters.)

GEORGE: Hello—sorry I'm late.

JUNE: Hello, George.

GEORGE: Hello. Hello, Rory.

RORY: Hello, George. We were just talking about last year when you missed your flight to Seattle.

GEORGE: Ah, yes—the stomach virus.

RORY: So you said.

GEORGE: What?

RORY: We were afraid you were going to miss the train. It's almost time for it to leave.

GEORGE: Ah—but in the subatomic world, time moves in both directions.

RORY: Do I look like a quark?

GEORGE: Actually, now that you mention it, old man—

JUNE: We should get on the train.

GEORGE: Ladies first.

RORY: Ladies? Where?

JUNE: Very funny.

GEORGE: No bouncy bouncy for you tonight, old boy.

(They get on the train. The men follow June as she looks for an empty carriage.)

JUNE: This one all right?

GEORGE: Perfectly fine. Whatever you like.

RORY: You know, George, there's such a thing as being too agreeable.

GEORGE: Good heavens. (to JUNE) Do you find me too agreeable?

JUNE: I think Rory is just feeling cranky—aren't you, darling?

GEORGE: Are you feeling cranky, old man? Miss your afternoon nap, did you?

RORY: All right, you two—stop it.

(There is an awkward pause followed by the sound of a train whistle. There is a slight jerk as the train begins to move.)

GEORGE: I'm glad we were able to get tickets to this play.

JUNE: I'm a little surprised it's still running.

GEORGE: Why?

JUNE: Well, how many people want to see a play about physics?

RORY: You should give people more credit, June.

(Pause.)

GEORGE: Oh, I have a joke for you. How many physicists does it take to screw in a light bulb?

JUNE: I don't know.

GEORGE: Well, it depends.

JUNE: On what?

GEORGE: On whether the light is a particle or a wave.

JUNE: Oh, that's a good one! I like that.

RORY: So how many if it's a particle?

GEORGE: That's it—that's the joke.

RORY: Oh. I see.

(There is another awkward pause.)

JUNE: I am dying for a cup of tea.

GEORGE: How very British of you.

RORY: (to George) Sometimes I think she's more English than I am.

JUNE: (to George) Would you like some?

GEORGE: Milk and sugar, please.

JUNE: Rory?

RORY: None for me, thank you.

(She leaves. There is a moment of awkward silence between the men.)

GEORGE: I quite enjoyed your lecture on M-theory.

RORY: Oh, thank you. Thanks very much.

GEORGE: I found it stimulating. Not my sort of thing, you know, but—

RORY: What do you mean "not your sort of thing"?

GEORGE: I'm a cosmologist, not a particle physicist.

RORY: Right.

GEORGE: I'm interested in the Big Picture.

RORY: Well, M-theory is kind of the Big Picture.

GEORGE: But it's all highly speculative, isn't it?

RORY: You mean I could be right, or I could be whistling through a straw up my ass.

GEORGE: A straw up your ass? What on earth does that mean?

RORY: It's just an expression.

GEORGE: I never heard it before.

RORY: You didn't grow up in Yorkshire.

GEORGE: Neither did you. In your lecture you described an extra dimension bordered on two sides by two membranes sitting in space—(he rips two sheets of paper from his notebook and holds them up)—like these two pieces of paper.

RORY: Right.

GEORGE: (Pointing to one of them) And we're living on this sheet.

RORY: Right. (Pointing to the other one) Or that one.

GEORGE: One of them, at any rate.

RORY: So what's the problem?

GEORGE: Well...it's a bit wonky—

RORY: It's no more wonky than your theory of everything in the universe being different vibrational modes of tiny strings.

GEORGE: Yes, but string theory—

RORY: You yourself have admitted that getting particle physics right was one of the problems of string theory.

GEORGE: I suppose—

RORY: So it sounds as though you could use my help.

GEORGE: Ah, yes—in search of the coveted Theory of Everything.

RORY: M-theory has at least as much chance of being right as string theory. Anyway, no theory can be proven conclusively – only unproven.

GEORGE: "I have heard the mermaids singing, each to each. I do not think that they will sing to me."

(June enters with tea. She hands one to George and sits down with hers.)

JUNE: Here we are. (imitating Rory) Nothing like a good cup of tea, eh?

RORY: He's quoting Eliot again.

JUNE: Oh, dear. "Prufrock" or "The Wasteland"?

RORY: "Prufrock."

GEORGE: Good for you, old man! And here I thought you knew nothing about poetry. June, did you ever hear the expression "a straw up your ass"?

RORY: That's "whistling through a straw up your ass."

JUNE: No, but what a charming image.

GEORGE: You mean in all your long years of marriage, Rory has never used that expression?

JUNE: No, never…although I…

GEORGE: What?

JUNE: I just had the oddest feeling of déjà vu.

GEORGE: Spooky.

RORY: I wonder when we have those feeling if it means we really have lived through it all before?

JUNE: What would be the probability of that?

GEORGE: Ah—probability is your specialty, isn't it?

(The blackboard appears. June moves over to it as the lights go down on the men.)

JUNE: Probability in quantum physics is different than probability in say, gambling. A better word for it might be "potential." (She draws an electron cloud.) For instance, an electron (she circles one electron) has the potential for behaving like a wave (she draws waves) and—when observed—a particle. (She draws a particle. As she does, Marie Curie and Newton wander in, pull up chairs and sit in the them, as though they were her "students." June does not interact with them, but they interact with each other.)

This deeply mysterious double nature of subatomic particles was the basis of the Heisenberg Uncertainty Principle, and was most dramatically illustrated in the famous "double slit" experiment.

(She reveals a diagram of the "double slit" experiment.)

NEWTON: (*sotto voce,* to Marie) I was the first *real* physicist—they knighted me, you know.

MARIE: So I heard.

JUNE: This experiment showed that a beam of electrons passing through a double slit would behave as a wave, creating a pattern on the opposite wall characteristic of wave interference.

NEWTON: (to Marie) Try to imagine the world before I came onto the scene—it was ruled by superstition!

MARIE: Shh!

JUNE: However, when a *single particle* was beamed at the slits, the same thing happened! How could this be? It seemed to defy all expectations of the way a particle "should" behave! Instead of choosing one slit or another, the electron *always* appeared to pass through *both* slits *at the same time,* as though it were a wave! But here's the really spooky part: at the moment it was "caught" by an observer, it would innocently throw up its hands, as it were, and appear to the observer as a particle!

NEWTON: (loudly) Ah—free will versus destiny!

MARIE: So then electrons have free will until we observe them?

JUNE: At that moment, we speak of the collapse of the *probability wave* of the electron. At any given moment there is only a *probability* of the electron being in one place or another. Observed, it becomes an individual particle—unobserved, it somehow maintains its wave function and behaves as a wave! Thus it has the ability—still not fully understood—to pass through both slits *at the same time.*

NEWTON: That's just wrong.

MARIE: Shh!

JUNE: So even though the electron would always "declare" itself as a particle the moment it was observed, it could not be forced to behave as a particle when travelling unobserved through space! So we cannot speak of an electron being both a wave *and* a particle, but of being *either* a wave *or* a particle at any given moment.

NEWTON: That's it—I'm leaving!

(He stalks off.)

JUNE: That means the moment you observe a particle, you change it forever. Therefore, you can never really observe the present. And if you can't even observe the present, how can you possibly predict the future?

MARIE: How indeed?

JUNE: But—

MARIE: Some questions have no answers.

(June smiles at her as the lights go up on the men.)

RORY: M-theory is no less reasonable than string theory!

GEORGE: But there's still the problem of quantum corrections and gravity!

(Newton appears; only George can see him. The lights go down on Rory.)

NEWTON: Gravity's really holding you all up, isn't it? Oh, that's a good one! Gravity—"holding you up"!

GEORGE: Actually, we could have predicted the expanding universe from your theory of gravity.

NEWTON: Yes, yes—but in my day everyone had too much invested in the idea of a static universe, dear boy. It seems every time an advance is made in cosmology we move farther and farther from the center of things.

GEORGE: "The center will not hold."

NEWTON: What?

GEORGE: It's a quote from a twentieth-century poet—Yeats.

NEWTON: He sounds Irish.

GEORGE: He was.

NEWTON: They're all Catholics...I invented the *world,* dear boy! My three laws of motion—along with my discovery of gravity, explained the movement of the cosmos! What is it you have now at the center of creation?

GEORGE: The Big Bang.

NEWTON: The "Big Bang" indeed—how vulgar! What is it they've turned my gravity into—curved space? (He shudders.)

GEORGE: Space-time, actually.

NEWTON: (disgusted) Space-time...What a mess.

(He leaves.)

GEORGE: Maybe. Maybe not.

(He turns back to the others.)

JUNE: Do you think there's even a remote probability we could all just enjoy this train ride?

GEORGE: As I said before, probability is your specialty.

JUNE: *Touché,* George. What a witty comeback.

RORY: He should have been a bloody stand-up comic.

GEORGE: I prefer to work sitting down.

RORY: (pointedly) Or lying down...

GEORGE: What?

RORY: Nothing.

JUNE: (taking the tea from Rory) How about some tea, then?

RORY: When did you start liking tea so much, darling?

JUNE: Oh, I've developed all sorts of tastes since I've been in England.

RORY: So what exactly is your problem with M-theory?

GEORGE: Oh, here we go again!

RORY: I don't see that it's so different from string theory! Membranes can be thought of as a series of strings—but in M-theory the strings are stretched into an extra dimension.

JUNE: Rory's right. String theory suggests we're all just vibrations floating around in space.

RORY: That's what all those damn mystics have been on about all these centuries: that we are all connected notes in some bloody cosmic symphony.

GEORGE: Ah—but who's plucking the strings?

RORY: No one's bloody plucking the strings, or moving the membranes, or anything—it's just undulating, all by itself, without any Hand of God or Great White Father in the Sky, or Universal Intelligence, if you're a Methodist. It's just sitting there like a bunch of seaweed waving about in the tide.

GEORGE: How do you know?

RORY: What?

GEORGE: How do you know there's no Universal Intelligence out there right now, watching over everything, making sure this train doesn't crash. Or making sure that it will.

RORY: I don't know. No one's ever proved it to me.

GEORGE: No one's ever really proved that atoms exist, and yet you believe it.

RORY: That's different. Experiments have shown—

GEORGE: Are you saying you're immune to faith?

RORY: No, not immune, I just—

JUNE: George, Rory has as much right not to believe as you do to believe.

RORY: Why do you have faith?

GEORGE: We all have faith—it just goes by different names.

RORY: Why do you *need* faith?

GEORGE: There's a—a faculty within me, a capacity—for faith, if you will. Don't you feel wonder and awe when you consider Nature? The workings of an atom—or, say, a bee hive?

RORY: Wonder and awe, maybe—but not faith.

GEORGE: Well, in me it manifests as faith. There's a place where reason and logical answers don't do the job anymore.

JUNE: A kind of dead end—?

GEORGE: Yes—where reason fails. Reason and logic fall away like the discarded, dried shell of a snake shedding its skin. Haven't you ever had that experience? Where you didn't wonder why something was, because something inside you *knew*?

RORY: God doesn't enter into it. I left all that long ago.

GEORGE: You were raised what—?

JUNE: Presbyterian.

RORY: Scottish mum and all that.

JUNE: So what happened?

RORY: When I was about ten, my cousins were visiting from America, and they told us about this television show about cave men—

JUNE: *The Flintstones*!

RORY: Right—*The Flintstones*. Well, one day in church I was reciting the Nicene Creed, feeling all cozy, like a good little Christian and everything. Then I started thinking about *The Flintstones,* so I tried replacing all the words of the Creed with "yabba dabba doo"—and it felt just the same. I lost my faith that day.

GEORGE: Oh, come on—it had to have been more than that.

RORY: Oh, certainly. I took a look around the world, and decided that: A) God probably doesn't exist, and B) If he does, he's a real shit. And what use is a god like that?

GEORGE: But we can't know His mind.

RORY: You know what? He's welcome to it. He's either a sadist or has a really sick sense of humor!

GEORGE: You sound angry at God.

RORY: He took my son from me, George! How do you *expect* me to feel? (Pause.) Christ...I need a cigarette!

(He leaves.)

GEORGE: I thought he quit.

JUNE: He did.

GEORGE: So that was all about David.

JUNE: He's having trouble accepting what happened.

GEORGE: Can't say I blame him.

(Pause.)

JUNE: Did you really say you missed the flight because of a *stomach* virus?

GEORGE: Yes. Why?

(She looks at him without responding.)

GEORGE: I couldn't very well say I was with you, could I?

JUNE: So I'm a stomach virus—

GEORGE: It was the first thing that occurred to me.

JUNE: Oh, George. You're a rotten liar.

GEORGE: Sorry.

JUNE: It just makes you more...lovable.

GEORGE: Lovable? Because I'm a rotten liar?

JUNE: (sighing) Poor Rory.

GEORGE: First I'm a rotten liar and now it's "poor Rory"?

JUNE: He's the only one who doesn't know what's going on.

GEORGE: How do you know? Maybe he does.

JUNE: Trust me—he has no idea. Do you think he'd leave us alone together like this if he did?

GEORGE: No, I suppose not. I would, though.

JUNE: Would what?

GEORGE: Leave the two of us alone.

JUNE: That's different. You're you and he's—Rory.

GEORGE: I see your point.

(Rory enters.)

RORY: Do you know that the penis of a humpback whale is twelve feet long?

GEORGE: Good God, Rory.

JUNE: What made you think of that?

RORY: I remembered what you said about the reason you married me. (to George) So if you laid us end to end, we still wouldn't be as long as the penis of a humpback whale.

JUNE: (overlapping)—as the penis of a humpback whale!

GEORGE: All right, Rory—I *get* it!

RORY: You're cranky. (to June) Why is he so cranky?

JUNE: Maybe he missed his afternoon nap.

RORY: Oh, we missed our nappies, did we?

GEORGE: Don't push your luck.

JUNE: I thought your lecture was brilliant, darling. Maybe M-theory will turn out to be the unified Theory of Everything.

GEORGE: Then you'll become famous, old man, just like Einstein. Oh, that's right—your hero is Max Planck.

RORY: You know, sometimes when I'm about to fall asleep, I see this long line of scientists who came before us...a glorious, ancient chain of truth seekers whose work stretches out across the centuries: Plato, Aristotle, Galileo, Newton, Poincaré—

GEORGE: Faraday, Thompson, Maxwell, Planck—

JUNE: Dirac, Einstein, Bohr, Pauli—

RORY: Schrödinger, the Curies, Heisenberg—

GEORGE: Fermi, Feynman, Gell-Mann—

JUNE: Hubble, Hawking, Ed Witten . . .

RORY: They are our Magellans, our Leif Ericksons...they've made our discoveries possible, just as we'll help make possible the discoveries of those who come after us. Isn't that one of the most beautiful things about humans—our curiosity about the world around us, our need to know where we came from?

JUNE: And where we're going.

RORY: Yes, that too.

JUNE: But are we *evolving*? Knowledge is just knowledge; it has no moral charge any more than a neutron has an electrical charge.

RORY: But isn't it just possible that science has advanced this far because it is *meant* to advance...and great geniuses are like mile markers on a path leading us ever closer to the ultimate truths of our world? And that these truths are given in code, through insight – in bits and pieces—to artists, writers, philosophers and mystics, like a bloody great divine jigsaw puzzle? And so we keep trying to reach into the eternal—to find expression for it in words, music or painting or philosophical insight—but it remains ever just outside our grasp. Just like in quantum physics, the more closely we try to describe something, the more elusive it remains. Like a scittering electron hit by a beam of photons, the mere act of observation sends our quarry fleeing from the lens of our inspection.

JUNE: Oh, Rory, I...I could fall in love with you all over again.

GEORGE: Steady on, June. (Getting up) I think I'll get a bit more tea. Rory, old man, if you don't mind my saying so, you're beginning to sound a little bit like me.

(He leaves. A blackboard appears Stage Right. The lights go down on George and June and up on the blackboard. Rory walks over to it and writes $E=mc^2$ on the blackboard as Marie Curie and Newton enter and sit, as though they were his "students.")

RORY: You all recognize this equation. Some of you may even understand it. It is the most famous mathematical statement in the world. But how many people know this one? (He writes $E=hu$ on the board.) "E," of course, is energy; "u" is the frequency of a wave of radiation, and this—(he circles the "h")—is Planck's Constant.

(Max Planck enters and stands next to Rory.)

PLANCK: (to Marie and Newton) I never set out to be a revolutionary, you know.

NEWTON: Tell me about it.

MARIE: Shh!

RORY: Planck's Constant. There are only a handful of constants in theoretical physics—

MARIE: The speed of light, for example.

NEWTON: Shh!

RORY: In the early 20th century, physicists faced a puzzle known somewhat dramatically as the Ultraviolet Catastrophe.

NEWTON: Oh, that *is* dramatic!

RORY: Classical physics predicted that a theoretical object that absorbs electromagnetic radiation perfectly—known as a "black body" because it reflects no radiation—should also *emit* radiation perfectly, equally at all frequencies.

(Planck rises and draws a box with waves of various lengths radiating into it.)

RORY: And since the number of possible frequencies was thought to be unlimited, that would mean that a black body would emit infinite energy.

(Planck draws waves leaving the box.)

PLANCK: But since there's no such thing as infinite energy, that was obviously impossible. In fact, findings showed that the emission of energy actually *slowed down* and eventually stopped at the shorter, or ultraviolet, wavelengths.

(Planck draws a ray of radiation with a very short wavelength stopping suddenly.)

No one knew how to describe this effect mathematically. And in physics, if you can't describe something with an equation, you can't begin to fully understand it.

RORY: Physicists were stuck. That is, until Max Planck came up with his constant, and this equation.

(Planck points to the equation on the board.)

Planck realized that radiation is actually emitted not in a continuous, smooth stream, but in discrete packets – or "quanta"—which is Latin for "how much," as in "quantity."

(Planck draws a circular "packet" moving through space.)

This effectively solved the Ultraviolet Catastrophe! The very short wavelengths simply aren't capable of emitting an entire "packet"—or "quanta"—of energy. Planck's Constant—and this deceptively simple equation – at last provided a formula for the puzzling relationship between a wave's frequency and its emitted energy. This opened the gate to quantum physics, and began the long scrabbly climb toward its now famous Uncertainty Principle. It was the first nail in the coffin of classical physics.

NEWTON: Aha—so it's all *your* fault!

PLANCK & MARIE: Shh!

RORY: This was, in its own way, as revolutionary a statement as Einstein's more famous equation. And yet...everyone knows about the great Albert Einstein, but how many people have heard of Max Planck?

NEWTON: (to Planck) I know just how you feel, old boy—(as Marie drags him offstage) We'll discuss this later.

(Rory and Planck are left alone.)

PLANCK: My constant was an act of desperation. I had struggled with the problem for years *und* now I was backed into a corner. On the surface, my solution made no sense.

RORY: But it has persisted—it's lasted a hundred years!

PLANCK: But when Einstein later used my equation to describe the photoelectric effect, *he* won the Nobel Prize for it!

RORY: But you—*you* were the first great pioneer of quantum mechanics!

PLANCK: Perhaps, but I'm not the one people remember.

RORY: Still, I envy you...you were so—dedicated. You were rejected time and time again, but you never gave up.

PLANCK: And spent the rest of my life trying vainly to backpedal away from the implications of my discovery—to make it fit with Newton *und* classical physics!

RORY: But you had surpassed Newton.

PLANCK: Not surpassed—*nein*. I was a classicist by training and temperament. It's ironic, you know, my discovery of the constant.

RORY: How so?

PLANCK: Because since then we've learned that nothing is constant…I lost my son too, you know.

RORY: Yes, I know. How did you—?

PLANCK: Science. And music. Or I suppose you could say physics, because music is based on physics too…it must be hard for you; everything is changing so fast in physics nowadays.

RORY: Equations have a kind of beauty about them, a purity…like a Bach partita.

PLANCK: I prefer the Romantics, actually.

RORY: Really?

PLANCK: Jah…Schubert, Beethoven, Brahms. I do like the *St. Matthew Passion*, however.

RORY: Interesting.

PLANCK: What?

RORY: It's George's favorite piece of music.

PLANCK: George?

RORY: A friend—a colleague—a rival of mine.

PLANCK: Ah. Your relationship is a double helix—two intertwining strands, like DNA.

RORY: How do you know about the discovery of DNA?

PLANCK: I try to keep up.

(A piano and cello appear.)

PLANCK: Shall we?

RORY: All right.

(They play, Planck at the piano and Rory on the cello. As they do, June walks over to the blackboard. She lectures to imaginary students.)

JUNE: Do you think you're a "good" person, whatever that means? (Listens for answer) More or less? Well, which is it, more or less? (Listens) Okay, then—more. So you're a "good" person. (She turns to another "student") How about you? (Listens) Okay, no doubt in your mind; you're a "good" person. Fine. Ever tell a lie? Steal anything? Cheat on your husband? (Pause) What were you at that moment? Were you a "good" person at the moment you were lying or stealing? (Listens) No? Okay. Does that negate your being a "good" person in general? No? But would you agree that at the moment you stole or lied or cheated, you were no longer a "good" person? You could go back to being one afterwards, but at that moment you gave up your status as "good."

Now pretend you're an electron. And you're happily zipping around as a wave—and then some grumpy physicist fires a photon at you, wanting to measure you—and at that moment you stop being a wave and become a particle. You can go *back* to being a wave later, but you can never be both things at once—just like you can go back to being a "good" person, but you can't be one at the moment you're being "bad."

Maybe the electron has a memory of itself as a wave at the moment we perceive it as a particle, but we can only see it for what it is at that moment in time. The more accurately you try to measure position, the less accurately you measure velocity. You have to use at least one quanta to measure a particle, which then moves the particle, thereby changing it.

(Rory stops playing and goes over to June.)

RORY: But what if you could observe something—or someone – without really changing them at all?

JUNE: You know as well as I do that's impossible.

RORY: But is it really? I mean, what if you come upon someone while they're sleeping—having an afternoon nap, say, and you just sit quietly and watch for a while. But they remain asleep so you haven't really affected them in any way.

JUNE: But you *have* affected them, even if you don't know in what way. The observer is *always* a participant.

RORY: There was an afternoon in early May when you had come in from the garden and I guess you'd laid down on the sofa for a quick rest before going back out again, but you'd fallen asleep and when I came into the room you didn't even stir, you just lay there on your back with one arm flung over your head, still wearing your green canvas gardening gloves, a bit of damp earth clinging to the fingers of the gloves, and I could smell the earth and it mingled with the scent that seems to emanate from the back of your neck; it's a spicy warm smell like gingerbread men baking, and I crept up and leaned over and just inhaled that scent for a while and you didn't move at all, not once, and there was a smudge of dirt on your nose and another one on your chin and your face looked pink and sunburned and there was a thin mist of sweat on your upper lip, and then I stole out of the room and went back to my study and when you came in later with a cup of tea I never mentioned it, watching you sleep like that. I just wanted to keep it as my secret – because then no one could take it from me.

JUNE: Why are you telling me now?

RORY: Maybe because it doesn't matter any more.

JUNE: Why not?

RORY: Because you're sleeping with George. (Pause.) Oh, June, don't you see? I *am* Schrödinger's cat! I don't exist until someone observes me . . . until someone looks inside the box where I am, I don't even exist as a probability. And by "someone," I mean you.

JUNE: Oh, Rory.

(George enters.)

GEORGE: Did I interrupt something?

RORY: No, nothing.

(June looks at him for a moment and then exits.)

GEORGE: What's wrong with her?

RORY: Far as I can make out, she's upset because she's sleeping with you.

(George is silent.)

RORY: What, you thought I didn't know? Why is it you both assumed I didn't know? Do you really think I'm that dim?

GEORGE: Look, old man—

RORY: No—no, *you* look! I don't even mind that much that you two are...I always thought she preferred you anyway. What I really mind is that you both seem to think that I'm so out of touch that I—

GEORGE: Look, old man, I'm frightfully sorry.

RORY: Don't apologize. For God's sake, don't apologize!

(There is a silence. George looks meditative.)

RORY: What?

GEORGE: I was thinking how Nature sends us enough trouble but we somehow have to invent our own anyway.

(George exits. June enters and goes to the blackboard as the lights go down on Rory. She lectures to imaginary students.)

JUNE: Gravity. The word is always associated with mass, weight—grave, gravitas; and yet of the four fundamental forces—electromagnetism, the strong and weak nuclear forces, and gravity—it is by far the weakest. And yet it is the one we experience most directly in our daily lives.

It is the reason we are stuck to the surface of our planet, rather than flying off into space like milkweed carried by a sudden gust of wind.

In the subatomic world there is an unavoidable unpredictability or randomness. Nothing in Nature can be predicted with utter certainty. Take the beta or alpha decay of radioactivity. Or the collapse of a building. Or the collapse of a life. If a plane collides with a solid mass—a building, say, there is always the chance that it will not collapse—but the probability is that it will, just as the wave function of an electron collapses when it is bombarded by a photon of light.

Life is therefore always a gamble.

So a building collapses, brought down by the weakest force of all—by what is really just a curve in the space-time fabric.

I felt a distortion of space-time that day…it was as though eternity was wrapped inside that moment.

(Lights up on Rory. June joins him.)

RORY: (to June) I don't really blame you, you know. I'd prefer George over me if I were you.

JUNE: (to Rory) Oh, Rory, don't you see? It's not about me preferring George—I just want to make time go backwards, so that I could have *been* there when it happened!

RORY: But why, June?

JUNE: Because maybe I could have done something! Maybe I could have saved him—saved our son—but now I'll never know!

RORY: Darling, I never blamed you for David's—

JUNE: I blame myself! Don't you see that? How can you not hate me for not being there with him?

RORY: But I would still love you, even if you *were* responsible for his death, which you weren't!

JUNE: But I can never relive that moment, never know what might have happened…

RORY But that's what life *is*! There's always that moment—

GEORGE: No larger than a lepton—

RORY: A neutrino—

GEORGE: A quark—

JUNE: A moment of decision—

RORY: To turn left or right—

GEORGE: To go now or wait—

JUNE: To pour another cup of coffee—

GEORGE: And therefore miss a bus—

RORY: Or a plane—

JUNE: Or a train…miss a ride that could change your life forever. And all because you decided to have an extra cup of coffee. Can it be that these moments are the structure of our lives?

RORY: A letter unanswered—

GEORGE: —a phone call ignored—

JUNE: —a tiny lump in your breast unexamined.

GEORGE: And so you miss the opportunity for reconciliation—

RORY: —the possibility of true love—

GEORGE: —the chance for a cure.

JUNE: Some of these moments are just the result of chance—

RORY: Accident.

GEORGE: Happenstance.

JUNE: Uncertainty. Everything is uncertain. It all goes back to Heisenberg.

RORY: Bloody Nazi.

GEORGE: But what if he wasn't? Not in his heart, anyway? What if he really did sabotage the Nazi effort to build a nuclear weapons reactor?

RORY: What if I'm the Queen Mother?

JUNE: We'll never really know.

RORY: Knowledge is an illusion.

GEORGE: Control is an illusion.

JUNE: Even objectivity is an illusion.

RORY: And so is God—or should I say the Father, the Son, and the Holy Ghost? The Holy Trinity...

(Max Planck appears.)

PLANCK: As in Trinity Site, where you developed the atomic bomb during World War II. Who came up with that name?

RORY: Fermi. Enrico Fermi.

PLANCK: What an odd name for a bomb laboratory.

(He stands behind the three of them, watching them.)

JUNE: And here we are, three scientists stuck in another Eternal Triangle—

RORY: Three blind mice—

GEORGE: Three thousand dead.

JUNE: Three thousand citizens of twenty-seven countries. And suddenly all the time you had to spend with someone is...spent.

(Isaac Newton appears.)

NEWTON: So, my boy, smash any good atoms lately?

GEORGE: Very funny. You must have been the life of the party four hundred years ago.

NEWTON: Oh, no—I was dreadfully uncomfortable around people. So I turned to science, which made sense to me; far less complex than human beings. The way most people can read other people, I could read Nature.

GEORGE: You had that gift.

NEWTON: Gift and a curse, my boy—a gift and a curse. I suppose you might say my curse was my gift. Like your fellow—what's his name—Hawkins?

GEORGE: Hawking. Stephen Hawking.

NEWTON: Yes. Gift and a curse, both at the same time. Great mind, betrayed by his body. But who knows? If he hadn't gotten that wretched disease, he might have spent all his time playing cricket or football or some such thing. We only have so much time, you know, here on this earth.

GEORGE: Do we?

NEWTON: What?

GEORGE: Only have so much time? Or is there something I don't know?

NEWTON: I can't say, my boy—I really can't say.

GEORGE: Can't say or won't say?

NEWTON: No, I really can't say. There's a lot they don't tell us over here, you know. Want some advice?

GEORGE: Yes, of course.

NEWTON: Live in the mystery. Get used to it. It's really not so bad.

GEORGE: No, I suppose not.

NEWTON: Besides, if I really did have something revelatory to say to you—some amazing bit of knowledge that explained everything, what would you do with that knowledge?

GEORGE: I don't know…tell other people, I suppose.

NEWTON: They would think you were mad! You wouldn't be able to prove any of it, you know. You'd be ostracized. You'd be just like that wretched Galileo, shut up in his tower for all those years. Bloody Italians. (Pause.) So you're really a Catholic?

GEORGE: Yes, I am.

NEWTON: "When the stars threw down their spears / And watered heaven with their tears:"

GEORGE: "Did he smile His work to see? / Did he who made the lamb make thee?"

NEWTON: You have to take the tiger with the lamb. Unity of opposites, old man.

(George turns back to June and Rory as Newton joins Planck.)

GEORGE: Rory—

RORY: (to George) We used to be friends once.

GEORGE: Yes.

RORY: We read physics together.

GEORGE: Played rugby together.

RORY: Chased girls together. What happened?

GEORGE: You caught one of the girls—except that I wanted her too.

RORY: But you had everything—looks, title, class, money, charm. Why couldn't I have June? Is that so much to ask?

GEORGE: Apparently so.

RORY: I could kill you.

GEORGE: I could kill myself.

RORY: I trusted you.

JUNE: Why are you talking about me as though I'm not here?

RORY: This is between George and me.

JUNE: So I have nothing to do with it?

RORY: Not really, no. (to George.) I trusted you.

GEORGE: I could have warned you not to do that.

RORY: But you didn't.

GEORGE: I trusted myself, but I should have known better.

JUNE: What about me? I betrayed you too.

RORY: (to June) Do you love me?

JUNE: Yes—when you're not driving me crazy.

GEORGE: Do you love me?

JUNE: Yes. When I'm not with Rory.

GEORGE: Absence makes the heart—

RORY: Wither and die.

GEORGE: So you can only love someone if he's right in front of you?

JUNE: No—yes—I don't *know* anymore! I'm too damaged to love anyone perfectly ever again.

(Marie Curie appears.)

MARIE: Which of us can do that? Love isn't about perfection—

(Max Planck appears.)

PLANCK: We all have to—how do you say?—muddle through as best we can—

(Newton appears.)

NEWTON: —doing the best you can with whatever you have to give.

JUNE: You both deserve better.

RORY: But it looks as though we're stuck with you—at least for the time being.

JUNE: The time being...that's all there ever is.

GEORGE: The only thing I can do is to ask for forgiveness.

RORY: If you were in my position, would you forgive you?

GEORGE: I don't know. I've never been in your position.

RORY: It will take time.

GEORGE: How much time?

RORY: That depends upon the position and movement of the observer.

JUNE: Well, we're all on the same train now.

RORY: Or so we imagine. We could be on three separate trains.

GEORGE: (with a glance at Newton) Depending on who's observing us.

JUNE: I'm having that feeling again...

RORY: What feeling?

JUNE: Déjà vu.

GEORGE: "These beauteous forms, / Through a long absence, have not been to me / As is a landscape to a blind man's eye"

NEWTON: If the man is blind, is the landscape really there?

(George turns back to the others as Newton, Curie, and Planck watch the following exchange.)

GEORGE: Me too, come to think of it.

RORY: Maybe this is our chance to—

JUNE: To what?

RORY: To make it all come out right this time. (Pause. To George) Your strings have to vibrate in more than three dimensions, right?

GEORGE: Right—in order to accommodate all the particles you particle physicists have discovered.

RORY: What if they vibrate in eleven dimensions?

JUNE: Like in M-theory?

GEORGE: Why?

RORY: To accommodate gravity.

JUNE: You have to accommodate gravity.

GEORGE: These membranes...do they undulate or are they stiff?

RORY: (irritated) They *undulate.*

GEORGE: Steady on, old man—I'm not trying to tap into your inferiority complex.

RORY: Good, because I don't have an inferiority complex.

PLANCK: Yes, you do.

RORY: How do you know?

PLANCK: I'm your hero, remember? I know everything about you.

GEORGE: So they undulate—like this? (He demonstrates with his hand.)

RORY: Yes. They ripple—

GEORGE: Like waves! So what if the membranes really *are* just extended strings—

JUNE: In the eleventh dimension!

RORY: Yes—*that's* what I've been on about!

GEORGE: Living inside a higher dimensional space.

RORY: Yes—and there's your "god," living in those higher dimensions, if you must.

GEORGE: (impatiently) Whatever.

JUNE: So these membranes are separated by—

RORY: A spatial dimension.

JUNE: Is there any reason they couldn't...collide?

RORY: No, I suppose not.

JUNE: What if two of the membranes collided—sort of like you and George?

RORY: And when they collide, the energy turns into heat and light and matter—

GEORGE: Producing the world as we know it!

RORY: The rippling would cause them to collide unevenly—

GEORGE: —accounting for the lumpiness of matter in the universe—

JUNE: So for billions of years, they don't touch each other.

GEORGE: Right! They're cold and dark and empty. Cut out all features that can't be observed—

JUNE: Occam's Razor—

RORY: The simplest theory is the best—

GEORGE: So then instead of the Big Bang—

RORY: —the universe was formed by colliding membranes!

JUNE: We've always *assumed* the Big Bang was a singularity—

GEORGE: So no more singularity. At higher energies, the three forces unite into a single force—

JUNE: —and at very high energies, gravity becomes as strong as the other three forces—

RORY: —and matter becomes simpler and more symmetrical!

GEORGE: The physics would still be totally different, of course—

JUNE: But it solves the problem of the Big Bang!

RORY: That would make string theory—

GEORGE: A manifestation of—

ALL: M-theory.

RORY: (dreamily) M-theory.

JUNE: The Mother of All Theories.

NEWTON: Theoretically, then, the membranes could collide again in the future.

PLANCK: Yes.

MARIE: And again—

NEWTON: Yes.

PLANCK: An infinite future of Big Bangs—

MARIE: And parallel universes.

GEORGE: There's more work to be done.

RORY: Oh, tons.

JUNE: The equations—

GEORGE: Comparing of notes—

JUNE: The convergence of cosmology and particle physics.

GEORGE: Of string theory and—

RORY &
GEORGE: M-theory.

RORY: So time and existence don't begin with the Big Bang.

JUNE: That means no true beginning and no end to time.

RORY: (to George) Is that a problem for you? I mean, the Catholic Church gave its blessing to the Big Bang—

GEORGE: To hell with the Catholic Church! What do they know about science? (Looking at his watch) Are we there yet?

JUNE: Well, we're somewhere.

RORY: On a train.

JUNE: To somewhere.

GEORGE: "Oh, do not ask, 'What is it?' / Let us go and make our visit."

RORY: Eliot?

GEORGE: (nodding) "Prufrock."

JUNE: "You do not have to be good. / You do not have to walk on your knees / for a hundred miles through the desert, repenting. / You only have to let the soft animal of your body / love what it loves. / Tell me about despair, yours, / and I will tell you mine. / Meanwhile the world goes on."

(The men look at her.)

JUNE: Mary Oliver.

GEORGE: (to Rory) Well? Have we solved anything?

RORY: Does it matter?

GEORGE: To me it does.

RORY: "Unwearied still, lover by lover, / They paddle in the cold, / Companionable streams or climb the air, / Their hearts have not grown old, / Passion or conquest, wander where they will / Attend upon them still."

(George and June look at him.)

RORY: Yeats. As in William Butler.

GEORGE: Well done, old man. Have we arrived at a companionable truce, then?

RORY: Until the next time.

GEORGE: What next time?

NEWTON: Oh, there's always a next time.

JUNE: (looking out the window) The train's slowing down. I think we're there.

RORY: Let's go, then.

GEORGE: Let's go.

JUNE: I think I preferred the getting there.

GEORGE: How do you know? You haven't seen the play yet.

RORY: Well—shall we?

GEORGE: After you.

RORY: No, please—after you.

GEORGE: I insist. I'll go first the next time.

(They leave. Newton, Planck, and Curie take their seats on the train. There is a pause.)

PLANCK: Well, here we are.

MARIE: Here we are.

NEWTON: Tea, anyone?

PLANCK: Why not?

MARIE: Good idea. (She gets up to get it.)

NEWTON: Oh, please—allow me. It's the least I can do; after all, I started this whole mess.

MARIE: Thank you. That's very kind.

PLANCK: (Looking out the window) Where are we headed, by the way?

MARIE: I'm sure we'll find out when we get there.

NEWTON: We might as well enjoy the ride.

PLANCK: Good advice—for an Englishman.

NEWTON: How ironic—coming from a German!

MARIE: Please—let's not start! What about that tea, eh?

NEWTON: Ah, yes—it'll just be a moment before the dining car opens again.

PLANCK: Or an eternity.

NEWTON: How will we know the difference?

MARIE: We won't.

PLANCK: Einstein's theory of special relativity states that each observer has his own idea of time.

NEWTON: Oh, Good Lord—not *him* again!

MARIE: And that there is no absolute time.

PLANCK: And each observer's point of view is equally valid—

NEWTON: All right—all *right!*

MARIE: So we might as well enjoy the ride.

(They all look out the window. There is the sound of a train whistle, and a slight jerk as the train begins to move.)

MARIE: It's very beautiful out there.

PLANCK: Yes, it is.

NEWTON: Yes. Yes, it's very beautiful.

(They all look out the window as the lights dim to black.)

Suggested Reading

Danielson, Dennis, ed. *The Book of the Cosmos: Imagining the Universe from Heraclitus to Hawking*. Cambridge, MA: Perseus, 2000.

Davies, P. C. W. and Julian R. Brown, eds. *Superstrings: A Theory of Everything?* Cambridge: Cambridge UP, 1988.

Feyerabend, Paul. *Against Method: Outline of an Anarchistic Theory of Knowledge*. London: Verso, 1993.

Greene, Brian. *The Elegant Universe: Superstrings, Hidden Dimensions, and the Quest for the Ultimate Theory*. New York: Random, 1999.

---. *The Fabric of the Cosmos: Space, Time, and the Texture of Reality*. New York: Random, 2004.

Kaku, Michio. *Hyperspace: A Scientific Odyssey Through Parallel Universes, Time Warps, and the 10th Dimension*. New York: Oxford UP, 1994.

---. *Parallel Worlds: A Journey Through Creation, Higher Dimensions, and the Future of the Cosmos*. New York: Doubleday, 2005.

Koyré, Alexandre. *From the Closed World to the Infinite Universe*. Baltimore, MD: Johns Hopkins UP, 1957.

Krauss, Lawrence. *Hiding in the Mirror: The Mysterious Allure of Extra Dimensions, from Plato to String Theory and Beyond*. New York: Viking, 2005.

---. *The Physics of Star Trek*. New York: Basic, 1995.

Lakoff, George and Mark Turner. *Philosophy in the Flesh: The Embodied Mind and its Challenge to Western Thought*. New York: Basic, 1999.

Lakoff, George and Rafael Núñez. *Where Mathematics Comes From: How the Embodied Mind Brings Mathematics Into Being*. New York: Basic, 2001.

Penrose, Roger. *The Road to Reality: A Complete Guide to the Laws of the Universe*. New York: Knopf, 2005.

Randall, Lisa. *Warped Passages: Unraveling the Mysteries of the Universe's Hidden Dimensions*. New York: Ecco, 2005.

Serres, Michel and Bruno Latour. *Conversations on Science, Culture, and Time*. Trans. Roxanne Lapidus. East Lansing, MI: U of Michigan P, 1995.

Smolin, Lee. *Three Roads to Quantum Gravity*. New York: Basic, 2001.

---. *The Trouble with Physics: The Rise of String Theory and the Fall of a Science, and What Comes Next*. New York: Houghton Mifflin, 2006.

Sokal, Alan and Jean Bricmont. *Fashionable Nonsense: Postmodern Intellectuals' Abuse of Science*. New York: Picador, 1998.

Susskind, Leonard. *The Cosmic Landscape: String Theory and the Illusion of Intelligent Design*. New York: Little, Brown, 2005.

Traweek, Sharon. *Beamtimes and Lifetimes: The World of High Energy Physicists*. Cambridge, MA: Harvard UP, 1988.

Woit, Peter. *Not Even Wrong: The Failure of String Theory and the Continuing Challenge to Unify the Laws of Physics*. New York: Basic, 2006.

Contributors

Sandy Beck began her corporeal existence in Massachusetts. After graduating from a sheltered girls' prep school, she moved to New York City where she studied figure painting at the Art Students League and Film Studies at Hunter College. Subsequent years included travel in Southeast Asia, South America, and Europe, plus an MFA in Creative Writing. Currently, Sandy is working on a PhD in English. She lives by the sea in Port Townsend, Washington.

Félix T. Bermúdez de Castro y Sorondo's professional experience encompasses art, design, and architecture. He has won several art and design awards and has enjoyed a successful career with experience ranging from business, site management, and architectural projects such as retail, exhibition and leisure, conversions, the refurbishment of listed historic fabrics to a new luxury holiday villa. His talents and multiple skills have been profitably and successfully used across the board in England and other countries.

Robert Borski has written two books about the fiction of Gene Wolfe, *Solar Labyrinth* and *The Long and the Short of It*, and lives in central Wisconsin. He believes his exposure to the Duncan Yo-Yo Craze in the early 1960s, along with a fascination for Cat's Cradles and the like, predisposed him to an acceptance of string theory, but at the same time denies he still has trouble tying his shoes.

Carole Buggé has five published novels, four novellas and a dozen or so short stories and poems. Winner of the Euphoria Poetry Competition and the Eve of St. Agnes Poetry Award, she is also the First Prize winner of the Maxim Mazumdar Playwriting Competition, the Chronogram Literary Fiction Prize, Jerry Jazz Musician Short Fiction Award, and the Jean Paiva Memorial Fiction award, which included an NEA grant to read her fiction and poetry at Lincoln Center.

Daniel Conover is a journalist and new-media maverick in Charleston, South Carolina. He blogs, cartoons, and makes films with his friends at <http://xark.typepad.com>.

It's a long story why college librarian **Lloyd Daub** of Milwaukee writes poetry using both male and female pseudonyms. A long story. Fortunately, there is Deb Kolodji's version: "There's the interesting question of

oino sakai and **Lucinda Borkenhagen** being two different literary personalities of the same person...Have they met? In which dimension?"

Diane Shipley DeCillis' poems and prose have appeared in *Nimrod International Journal, Rattle, William & Mary Review, Gastronomica, Connecticut Review, South Dakota Review, Puerto del Sol, Poet Lore, Sanskrit, Slipstream, Phoebe* and other journals. She won the *Crucible* Poetry Prize 2005 and the 2005 *MacGuffin* National Poet Hunt. She owns an award-wining gallery in Southfield, Michigan and is co-editor of *Mona Poetica,* an anthology of poetry on the Mona Lisa published by *Mayapple Press.*

Lindsay A.S. Félix earned her BA in English from Mary Washington College in 1998 and MFA in Creative Writing from George Mason University in 2007. She teaches English composition at Northern Virginia Community College, works full time as a business analyst, and is a poet at night and on weekends. Lindsay married her fabulous husband, Marvin, in 2004 and has two wagging dogs, Max and Shamone.

Linda Nemec Foster is the author of seven books of poetry including *Amber Necklace from Gdansk* (finalist for the Ohio Book Award) and *Listen to the Landscape* (finalist for the Michigan Notable Book Award). Her poems have been published in over 250 literary journals (e.g. *The Georgia Review, New American Writing,* and *North American Review*), translated in Europe, and produced for the stage. She is the founder of the Contemporary Writers Series at Aquinas College in Grand Rapids, Michigan.

Cherryl E. Garner manages a small law office in South Carolina but has her deepest roots in Alabama, the inspiration for much of her poetry. Her poetic interest is in exploring family, landscape, the common and the divine in any forms that seem best to fit, and she waits patiently for an acceptable theory of everything. In the past year, her work has been included in *IBPC, The Rose & Thorn,* and *The Petigru Review.*

Paul Ginsparg is Professor of Physics and Computing & Information Science at Cornell University and is widely known for his development of the ArXiv.org e-print archive. He has published numerous papers in the areas of quantum field theory, string theory, conformal field theory, and quantum gravity. A Fellow of the American Physical Society, he has won many awards, including a MacArthur Fellowship in 2002, the PAM (physics astronomy math) Award from the Special Libraries Association, the Council of Science Editors Award for Meritorious Achievement, and the Paul Evans Peters Award.

Sheldon Glashow has done seminal research in the fields of elementary particle physics and cosmology. He played a key role in unifying the weak and electromagnetic forces and in creating today's successful Standard Model, for which he won, along with Steven Weinberg and Abdus Salam, the Nobel Prize for Physics in 1979. He is the Arthur G.B. Metcalf Professor of Physics at Brown University and author of some 300 research papers and three books: *Interactions* (with Ben Bova, 1988), *The Charm of Physics* (1990), and *From Alchemy to Quarks* (1993).

Andrea Gradidge, she who might be obeyed, in a dimension somewhere, has relocated in space from the UK to British Columbia, Canada. Her son, Benjamin, studies computer science, and her daughter, Jennie, emigrated back to England with husband Kevin and their boys. Andrea strings words into minimalist poems and also does some computer illustration.

Lauren Gunderson is a resident of New York City by way of Atlanta, earning her BA at Emory University. A nationally commissioned and produced playwright, as well as a short story author and poet, Lauren is currently pursuing her MFA in Dramatic Writing at Tisch School of the Arts at New York University. She teaches and speaks around the world on the intersection of science and theatre, as well as arts as activism. <http://laurengunderson.com>

Kathleen M. Heideman is a fellow of the National Science Foundation's Antarctic Artists & Writers Program. Her poems appear in the anthology *33 Minnesota Poets*, chapbooks (*Explaining Pictures to a Dead Hare; She Used to Have Some Cows; TimeUponOnce*), and various literary magazines — *Three Candles, Exquisite Corpse, Cream City Review, Water-Stone*, etc. Her poem "Woman in a Barrel, About to Go Over Niagara Falls" was nominated for a Pushcart. She lives in Minnesota.

Tania Hershman <http://taniahershman.com> is a former science journalist living in Jerusalem, Israel, and a founding member of the Fiction Workhouse online writing collective. Her short and very short stories have been published in *Cafe Irreal*, the *Hiss Quarterly*, *Front&Centre*, *Vestal Review*, *Steel City Review*, *Creating Reality*, *Entelechy Review*, the *Steel City Review*, *Riptide*, and *Transmission*, and the *Wonderwall* and *Ideas Above Our Station* anthologies from Route. She has had three stories broadcast on

BBC Radio 4. Tania's first collection, *The White Road and Other Stories,* will be published by Salt Publishing in 2008. Tania is the editor of *The Short Review* <http://theshortreview.com>, a site dedicated to reviewing short story collections and anthologies.

Brenda Hillman has published seven collections of poetry, all from Wesleyan University Press, the most recent of which are *Cascadia* (2001), and *Pieces of Air in the Epic* (2005), which won the William Carlos Williams Prize for Poetry. Hillman serves on the faculty of Saint Mary's College in Moraga, California, where she is the Olivia Filippi Professor of Poetry. She is involved in non-violent activism with Code Pink in the San Francisco Bay Area.

Heather Holliger is a teacher, poet, and activist in the San Francisco Bay area. She holds an MFA in creative writing from George Mason University and currently teaches writing and literature at Ohlone and Chabot Colleges. She is interested in the connections between poetry and social change and has served on the boards of several community-based poetry and spoken word organizations. She plans one day to found an artist community with fellow activists.

Daniel Hudon, originally from Canada, teaches natural science at Boston University. His first book, *The Bluffer's Guide to the Cosmos,* will be published in 2008 by Oval Books (UK). He has published more than two dozen travel stories in literary magazines, including, most recently, in *Bayou Magazine*. He is working on a series of short stories like the one in the present volume. He lives in Somerville, Massachusetts.

David Hurst's poetry has appeared in the online and print journals *Arsenic Lobster, In the Grove,* and *hardpan*. He feels the congruence of science and art is no accident; poetry walks that line merging code with rhythm, idiom and nuance. David received his MFA in Creative Writing at CSU, Fresno and teaches English at College of the Sequoias in Visalia, CA. He lives in California's Central Valley with his wife, several children, and various animals.

Colette Inez has authored nine books of poetry and won Guggenheim, Rockefeller and two NEA fellowships. Her latest collection *Spinoza Doesn't Come Here Anymore* has been released by Melville House Books, followed by her memoir *The Secret of M. Dulong* from the University of Wisconsin Press. A visiting professor formerly at Cornell, Ohio, Bucknell and Colgate

Universities, she is widely anthologized, teaches in Columbia University's Undergraduate Program, and has appeared on public radio and TV.

N.K. Jemisin is a writer of science fiction, dark fantasy, and unclassifiables like her contribution to this anthology. Her work has appeared in *Strange Horizons, Ideomancer, Escape Pod,* and on the 2006 Recommended Reading list from the Carl Brandon Society. She is fascinated by singularities, quantum or otherwise — including New York, where she now lives.

Jeff P. Jones's poetry chapbook, *Stratus Opacus,* is forthcoming from Main Street Rag Publishing. His poems are recently in *Blood Orange Review, Hawai'i Pacific Review, Puerto del Sol,* and elsewhere. He teaches writing at the University of Idaho.

Michio Kaku holds the Henry Semat Chair in Theoretical Physics at the City University of New York. He is the cofounder of string field theory and the author of several widely acclaimed books, including *Visions, Beyond Einstein, Hyperspace,* and most recently *Parallel Worlds*. He hosts a nationally syndicated radio science program. His website is <http://mkaku.org>.

Christine Klocek-Lim lives in Pennsylvania. Her poems have appeared in *Nimrod, The Pedestal Magazine, OCHO #5, About.com*, the *Quarterly Journal of Ideology,* and elsewhere. In 2006, her work was selected as a finalist for *Nimrod's* Pablo Neruda Prize for Poetry. She is editor of the online journal, *Autumn Sky Poetry,* and serves as Site Administrator for The Academy of American Poets' online discussion forum at *Poets.org*. Her website is <http://novembersky.com>.

Cleo Fellers Kocol proceeded from being a writer of fiction, non-fiction, and drama to poetry. *The Sacramento Bee* publishes her column about poets and poetry – also found online. Her poetry appears in *Poetry Now, Mobius,* and other journals, and was choreographed, set to music, and danced at the Palace of the Legion of Honor in San Francisco. She's also presented her poetry and drama at the Steve Allen Theater in Los Angeles, California. Kocol is 81 years old.

Deborah P. Kolodji works in information technology to fund other dimensions of her life, where she sometimes creates strings of haiku with other poets. Shunning the use of colorful pseudonyms, she is the president of the Science Fiction Poetry Association and a member of the Haiku

Society of America. Her work has appeared in a variety of places including *Modern Haiku, Strange Horizons, Frogpond, Eclectica, Poetic Diversity, Dreams and Nightmares*, and *St. Anthony Messenger Magazine*.

David Kopaska-Merkel describes rocks for the State of Alabama and publishes *Dreams and Nightmares*, a magazine of science fiction and fantasy poetry. In 2006, a collaboration with Kendall Evans won the Rhysling Award of the Science Fiction Poetry Association, long-poem category. Recent flash fiction can be found at <http://dailycabal.com>. Chapbooks are available through <http://genremall.com> and <http://spechouseofpoetry.com>. David lives in Alabama with artists and furry layabouts; learn more here <http://dreamsandnightmares.interstellardustmites.com>.

Dr. Kristine Larsen is Professor of Physics and Astronomy at Central Connecticut State University where she has been on the faculty since 1989. The author of *Stephen Hawking: A Biography, Cosmology 101,* and numerous articles on the history of women in astronomy and astronomy education, Larsen has also published and presented various scholarly works on the astronomical motifs and motivations in the works of J.R.R. Tolkien.

Elissa Malcohn edited *Star*Line* (Science Fiction Poetry Association) from 1986 to 1988. Her work has appeared in *Amazing, Asimov's, Full Spectrum* (Bantam), *Tales of the Unanticipated,* and elsewhere. She was a 1985 John W. Campbell Award finalist and a four-time Rhysling Award nominee. *Covenant,* the first volume in her *Deviations* series, is available from Aisling Press. Forthcoming fiction includes work in *Electric Velocipede* and *Helix*. See her website (search for "Malcohn's World") for more information.

Michelle Morgan's work appears in numerous journals, including *parva sed apta, Arsenic Lobster, JMWW, Salt River Review, The Aurorean, Plain Spoke, Pemmican, Ruined Music, Wicked Alice, Marginalia, Blood Lotus, Alimentum & New Verse News*, and in over half a dozen anthologies. She is editor of the online literary & arts journal *panamowa*: a new lit order, and she is also a graduate student in the R. Kelly School of Disembodied Metaphorics.

Dave Morrison, a high school graduate and above-average guitar player, has published two novels and two collections of poetry. Visit him at <http://dave--morrison.com>.

Conceived originally in Kiev, Ukraine, **Alex Nodopaka** was first exhibited by accident in Vladivostock, Russia. Then he finger-painted in Austria, studied tongue-in-cheek at the École des Beaux-Arts, Casablanca, Morocco. Now he's back to doodling with crayons but on human canvases. Full time artist, art critic, and poet, he still dreams of silent acting in an IFC or Sundance movie.

James O'Hern is the author of *Honoring the Stones* published by Curbstone Press. Raised on a ranch in South Texas, he was mentored by a Mexican Indian who taught him native ways. Thus, he became profoundly aware of how progress erases the collective memory of indigenous people. After 35 years as an investment banker, he is committed to honor the loss of cultural identity through language and poetry. He lives in New York City. <ohern_j@msn.com>

Lynn Pattison's work has appeared in: *The Notre Dame Review, Rhino, The Dunes Review, Heliotrope, The MacGuffin* and *Poetry East*. Besides her chapbooks, *tesla's daughter* and *Walking Back the Cat*, she is the author of the book, *Light That Sounds Like Breaking*. Her poems have been anthologized in several collections and have twice been nominated for the Pushcart Prize. She imagines that in another dimension she might really understand quantum physics.

Joseph Radke's poems have appeared in *Boulevard, Versal, Poetry East, Natural Bridge*, and several other journals. He is the recipient of an Academy of American Poets prize. He lives in Milwaukee where he teaches writing and works on *The Cream City Review*. *Salt&Sand*, his poetry manuscript, is seeking a publisher.

Michael Ricciardi is naturalist, teacher, writer/poet, designer, and multimedia artist living in Seattle, Washington. He is also a former professional illusionist. As a science writer/reporter, Michael has interviewed several renowned scientists, including physicists Brian Greene (Columbia), Paul Steinhardt (Princeton), and Nobel Laureate Ilya Prigogine. Ricciardi is also an award winning, internationally screened video artist, and has received several grants for his artistic projects, including a Paul G. Allen Foundation for the Arts grant (2002, 2003).

Adam Roberts was born two thirds of the way through the last century. He is currently Professor of Nineteenth-century Literature at Royal Holloway, University of London. Perhaps surprisingly, given that fact, he is also the author of a number of science fiction novels, the most recent of which are: *Gradisil* (Gollancz, 2006, which was shortlisted for the Arthur C. Clarke award), *Splinter* (Solaris 2007), *Land of the Headless* (Gollancz 2007) and *Swiftly* (Gollancz 2008).

Bruce Holland Rogers is an American writer living temporarily in London. His fiction has been translated into 23 languages and has won a Pushcart Prize, two Nebulas, and two World Fantasy Awards. His most recent collection is *The Keyhole Opera*, and more of his stories are available at <http://shortshortshort.com> along with information about an e-mail subscription service to his short-short stories. He teaches fiction writing for the Whidbey Writers Workshop low-residency MFA program.

Mary Margaret Serpento (a.k.a. *.mms) writes SF short poetry in English and French, and is a figment of her own imagination.

scifaijin Celtic and bilingual—
   a cat reborn as human
   this turn (of the wheel)

Beret Skorpen-Tifft writes poetry and fiction. She lives in South Portland Maine with her husband and two children. She received her B.A from Hampshire College and her MFA in fiction from Vermont College. Her work has appeared in *The Louisville Review, Passages North, The Red Owl, The Sow's Ear, The Sun, Maine Things Considered, The Maine Times, and The Bangor Daily News.* She is a long distance runner, completing 7 marathons.

Jarvis Slacks attends the University of North Carolina at Wilmington, finishing his masters degree in Creative Writing. He also writes a weekly nightlife column, *Good Evening*, for the *Star News* and is working on his first novel. He likes cookies and ice cream.

Elaine Terranova's most recent book of poems is *Not To: New and Selected Poems*. Other books include *The Cult of the Right Hand*, for which she won the 1990 Walt Whitman Award of the Academy of American Poets, *Damages*, and *The Dog's Heart*. Her translation of Euripides' *Iphigenia at Aulis* is part of the Penn Greek Drama Series. She was named a Pew Fellow in the Arts and has received an NEA fellowship.

Wendy Vardaman, Madison, Wisconsin, has a PhD in English from University of Pennsylvania, as well as a BS in Civil Engineering from

Cornell University. Her poems, reviews, and interviews have appeared in various anthologies and journals, including *Poet Lore, Main Street Rag, Nerve Cowboy, qarrtsiluni, Free Verse,* Pivot, *Wisconsin People & Ideas, Women's Review of Books* and *Portland Review.*

Cecilia Vicuña is a poet and artist born in Chile. She performs and exhibits her work widely in Europe, Latin America, and the US. The author of 16 books, her work has been translated into several languages. Among them: *QUIPOem,* Wesleyan University Press, 1997, and *Spit Temple,* forthcoming by Factory School Press, 2008. The co-author of *500 Years of Latin American Poetry* for Oxford University Press, she lives in New York and Chile. <http://ceciliavicuna.org>

Peter Woit is a Lecturer in the mathematics department at Columbia University and a researcher in the field of mathematical physics. He holds a doctoral degree in theoretical particle physics from Princeton and writes a blog, *Not Even Wrong,* which often takes a critical view of string theory. His book of the same name was published in 2006 in the US and the UK and has recently appeared in Italian and French translations.

Susan Zwinger has published seventy poems and five books of natural history, including *The Last Wild Edge, Stalking the Ice Dragon, Still Wild, Always Wild,* a Sierra Club book. She teaches for the Whidbey Writers Workshop Master of Fine Arts in creative nonfiction, keeps illustrated journals, and leads natural history and creative writing workshops across the West. Zwinger has an MFA in Poetry from University of Iowa in 1971. She lives on an island where she translates from eagle to English daily.

About the Editors

Sean Miller is finishing up a PhD in English at the University of London. His topic is "the cultural currency of string theory." Sean has published two novels and poetry in journals. Raised in Connecticut, he's travelled extensively, living and working in China, Japan, Spain, and India.

Shveta Verma has two masters, one in literature, the other in education. She has taught English in New York City high schools and is interested in postcolonialism and cultural anthropology. Her family emigrated to the U.S. from the Punjab when she was two.

Sean and Shveta met during a Masters course at Birkbeck College in London in 2003. In 2007 they got married in Maui and honeymooned in Peru.

About Scriblerus Press

Scriblerus Press is the publishing arm of Banyan Institute, a 501(c)(3) tax-exempt nonprofit corporation headquartered in New York City. Its mission is to nurture and promote eclectic literary works and writers—without too much concern for mass-market viability. The house operates leanly, under the fashion radar—and with an eye to speaking truth to power. Its pretense is to be a small, however provisional, antidote to the omnipresent hegemony of consumerism. <http://scriblerus.net>

Printed in the United States
112255LV00003B/73/P

9 780980 211405